Global Institutions and Social Knowledge

Politics, Science, and the Environment
Peter M. Haas, Sheila Jasanoff, and Gene Rochlin, editors

Peter Dauvergne, *Shadows in the Forest: Japan and the Politics of Timber in Southeast Asia*

Peter Cebon, Urs Dahinden, Huw Davies, Dieter M. Imboden, and Carlo C. Jaeger, eds., *Views from the Alps: Regional Perspectives on Climate Change*

Clark C. Gibson, Margaret A. McKean, and Elinor Ostrom, eds., *People and Forests: Communities, Institutions, and Governance*

The Social Learning Group, *Learning to Manage Global Environmental Risks*. Volume 1: *A Comparative History of Social Responses to Climate Change, Ozone Depletion, and Acid Rain.* Volume 2: *A Functional Analysis of Social Responses to Climate Change, Ozone Depletion, and Acid Rain*

Clark Miller and Paul N. Edwards, eds., *Changing the Atmosphere: Expert Knowledge and Environmental Governance*

Craig W. Thomas, *Bureaucratic Landscapes: Interagency Cooperation and the Preservation of Biodiversity*

Nives Dolšak and Elinor Ostrom, eds., *The Commons in the New Millennium: Challenges and Adaptation*

Kenneth E. Wilkening, *Acid Rain Science and Politics in Japan: A History of Knowledge and Action toward Sustainability*

Virginia M. Walsh, *Global Institutions and Social Knowledge: Generating Research at the Scripps Institution and the Inter-American Tropical Tuna Commission, 1900s–1990s*

Global Institutions and Social Knowledge

Generating Research at the Scripps Institution and the Inter-American Tropical Tuna Commission, 1900s–1990s

Virginia M. Walsh

The MIT Press
Cambridge, Massachusetts
London, England

This book was set in Sabon by Achorn Graphic Services.

Library of Congress Cataloging-in-Publication Data

Walsh, Virginia M.

Global institutions and social knowledge : generating research at the Scripps Institution and the Inter-American Tropical Tuna Commission 1900s–1990s / Virginia M. Walsh.

p. cm. — (Politics, science, and the environment)

Includes bibliographical references (p.).

ISBN 978-0-262-23236-4 (hc. : alk.paper)—ISBN 978-0-262-73167-6 (pbk. : alk. paper)

1. Marine sciences—Research—History—20th century. 2. Research institutes—Social aspects. 3. Scripps Institution of Oceanography. 4. Inter-American Tropical Tuna Commission. I. Scripps Institution of Oceanography. II. Inter-American Tropical Tuna Commission. III. Title.

GC57.W25 2004
306.4′2—dc22

2003064743

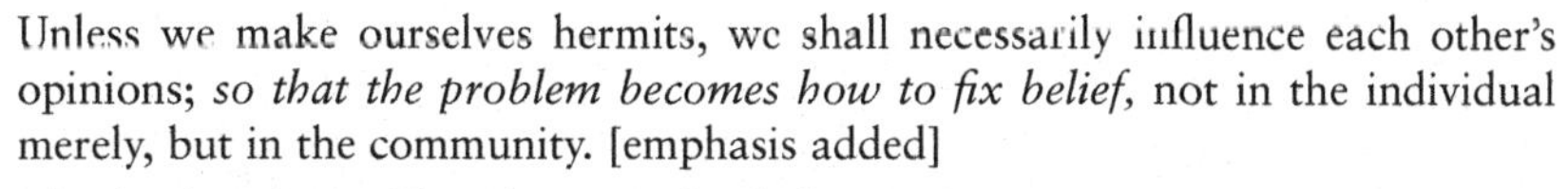

Unless we make ourselves hermits, we shall necessarily influence each other's opinions; *so that the problem becomes how to fix belief,* not in the individual merely, but in the community. [emphasis added]

Charles S. Peirce, "The Fixation of Belief," 1877

Contents

A Word to Readers

Virginia Walsh died of Hodgkins lymphoma before she could complete work on this book. At the time of her death, Virginia was about to embark on a final round of revisions. She had planned extensive revisions to further clarify and strengthen the argument and text. She had also planned to spell out in greater detail some of the implications of the approach to institutions and knowledge developed in the book. As her husband, I had the opportunity to have many discussions with Virginia about the book, and I have attempted to carry out many of the revisions she had planned in service of clarifying the argument and text. Lacking the requisite knowledge and expertise, and not wishing to put words in Virginia's mouth, I have not attempted to incorporate the revisions of content she had planned. However, I have added two paragraphs to the final chapter. These paragraphs contain a couple of ideas that I know Virginia felt were important consequences of the approach developed in this book. I have no doubt that she would have developed these points with greater subtlety, depth, and nuance; however, I felt a clumsy presentation of these points was better than none at all. These paragraphs have been identified in the notes so that readers will know they were added by me.

As I am unable to acknowledge all those who helped Virginia in writing this book, I hope that those who helped her in this process understand that their contributions are very much appreciated. I would be remiss, however, if I did not thank Oran Young, Raimo Tuomela, Peter Haas, John Odell, Ron Mitchell, and Clay Morgan for their comments, criticisms, encouragement, and support at various stages of the writing process. It is important also to acknowledge a research and writing grant from the MacArthur Foundation and a faculty fellowship leave from

Rutgers University, both of which provided Virginia with time to work on this book. Finally, I would like to thank the Department of Political Science at Rutgers–Newark for closing ranks during Virginia's illness. Their support allowed Virginia to continue to work on the book while she was undergoing treatment.

Virginia was excited about the prospect of extending and further developing some of the ideas and tools contained in this book and applying them to other problems concerning the environment and international relations more generally. I hope some readers are similarly excited and take up the challenge.

Sandeep Prasada
Department of Psychology
Hunter College, CUNY

Foreword

In this well-argued and thought-provoking study, Virginia Walsh makes an important contribution to our understanding of the roles that institutions play in international society. Many students of institutional arrangements—or regimes, as they are often called—have argued that knowledge is a driving force in the processes involved in the formation of specific regimes. But Walsh turns the causal arrow underlying such arguments around to consider the roles that institutions themselves play in the production and dissemination of knowledge. Focusing on the performance of regimes dealing with marine resources, and especially marine fisheries, she examines the operation of specific arrangements in depth and contributes to our knowledge of what others have termed the generative roles of environmental regimes. We have known for some time that regimes can make a difference as determinants of the way issues are framed and how we think about them, whether or not they prove effective in solving or managing the problems leading to their creation. But how exactly do institutions produce such results, and how significant is this element in the operation of regimes? These are the questions that Walsh tackles both analytically and empirically in this book.

Not content with measures of association, Walsh begins by asking questions about the causal mechanisms through which regimes contribute to the growth of knowledge. As a point of departure, she makes a persuasive case that beliefs—and socially accepted beliefs in particular—matter in international politics. This leads to a consideration of the various ways in which groups fix important beliefs. The use of the term *fix* in this connection is intriguing. It refers to social processes resembling the biophysical processes we have in mind in speaking of nitrogen fixation in ecosystems. The argument here is that institutions constitute key elements in the processes through which social groups generate ideas about major issues and proceed to fix social beliefs about them.

Having set the stage regarding the importance of such processes, Walsh proceeds to introduce three specific mechanisms through which groups fix beliefs: the positional fix, the statutory fix, and the committee fix. The positional fix centers on the actions of key individuals who are able to use their positions or roles to influence the beliefs held by members of their groups. The statutory fix refers to a process in which beliefs gain acceptance as a consequence of being embedded in formal or informal rules. For its part, the committee fix arises from efforts to arrive at consensus among scientists or groups of experts who meet regularly and act as a committee.

To move beyond the conceptual level in examining these processes, Walsh proceeds to explore them in detail through case studies of the evolution of the research programs of the Scripps Institution of Oceanography and the role of knowledge in the management practices of the regime created under the terms of the Inter-American Tropical Tuna Convention (IATTC). The result is a reciprocal relationship in which the mechanisms through which beliefs are fixed provide a well-defined road map for the case studies, and the case studies provide a rich empirical setting in which to flesh out the different fixes and explore their operation through a sustained assessment of the performance of real regimes.

To her credit, Walsh also recognizes the importance of dealing with counterfactuals in studying the consequences of institutional arrangements. To this end, she engages in a systematic effort to determine whether there are rival hypotheses or alternative explanations that can do as well or better in accounting for the evolution of knowledge about marine systems.

A neorealist account, for instance, would take the view that the development of beliefs is shaped by fluctuations in the distribution of power, so that changes in beliefs arising from the operation of regimes as such are epiphenomenal. Such an interpretation suggests that research conducted at Scripps should have turned toward defense-related or security issues with the rise of the United States to great power status from the 1890s onward. Yet nothing of the kind actually occurred. Despite major changes in the distribution of power, Scripps did not become deeply involved in defense-related marine research until after World War II. Similarly, an interest-based account would suggest that knowledge creation under the auspices of IATTC should reflect the interests of organized and influential actors, such as the major firms engaged in the tuna fisheries. Yet during the life of the IATTC—the period following 1950—beliefs about the relevant marine ecosystems have changed considerably despite the fact that interests in the tuna fisheries have not undergone any fundamental changes. These efforts to address key counterfactuals do not yield definitive evidence regarding the roles institutions play in the creation of knowledge, yet they do contribute significantly to the credibility of Walsh's claims about the roles institutions play in the processes through which beliefs arise and become socially accepted.

A particularly attractive feature of this account is Walsh's decision to focus on the role of ideas or cognitive processes without embracing the more extreme versions of social constructivism. Her entire argument addresses ideas and their connections to social institutions. But she concentrates on formulating and testing hypotheses in a manner that is fully compatible with the analytic practices of the new institutionalism. This is a significant contribution in a field of study that is increasingly divided by methodological and even epistemological disagreements between those who work within the mainstream of neoliberal institutionalism and those who criticize this work from a constructivist point of view.

The logical next steps in research on the links between institutions and knowledge will feature efforts to determine the relative weights of the processes Walsh describes, on the one hand, and a number of other determinants of the effectiveness of institutions, on the other, as well as to sort out the conditions under which the various types of fix come into play. There is nothing in Walsh's work that denies the importance of other drivers (for example, power and interests) as sources of influence in the operation of regimes. We are not faced with either/or propositions in this realm. Rather, Walsh's work enhances the toolkit available to those seeking to understand the impacts of regimes and helps to set the agenda for the next stage in this field of inquiry.

Sadly, Virginia herself will not be able to participate in this important effort. Her premature death at the age of 37 in January 2003 has cut short a career that had already produced important research findings and that promised to yield even more significant results in the future. This is a tragedy of the first order. Even so, her scientific legacy is likely to prove lasting. In a short period of time, she produced insights from which we can all benefit and pointed the way toward a line of inquiry that promises to yield cutting-edge results for some time to come.

Oran R. Young
University of California–Santa Barbara

Preface

This study began as a tightly focused investigation of the Inter-American Tropical Tuna Commission (IATTC), within the general framework on neoliberalism and negotiation theory. It soon became obvious to me that the IATTC was doing much more than facilitating the exchange of information among its contracting parties. The organization was generating original research and using it to regulate the fishery. Drawing upon background in international relations and economics, I had expected to see the organization act as a neutral broker, quietly urging fishing states to implement and enforce regulations.

This occurred, of course. But what surprised me was the extent to which the organization's scientific staff was actively generating new knowledge, not simply transmitting information generated by the parties themselves.

If the IATTC used its formal treaty rules to guide generation of new knowledge, I wondered, what could this tell us more generally about institutions and knowledge? (In other words, in the words of a skeptical adviser, "Who cares about this little fishery regime?") In principle, the generative function of institutions should be visible not only in specific treaty regimes ("the very small") but also in meta-regimes like sovereignty or property rights ("the very large").

To explore the effects of changes in these meta-institutions required going beyond my background in neoliberalism and bargaining theory. Is it possible to specify sovereignty in institutional terms, and how would one specify change over time? How could "knowledge" be captured as a concept subject to focused, comparative study? What would it mean for the relationship between the two to be "causal"? Respected colleagues encouraged me to continue asking "big questions" (even as I could hear the tenure clock ticking).

Although a number of my institutionalist colleagues were skeptical—neoliberals saw me as constructivist, and constructivists saw me as neoliberal—I found that much of the conceptual groundwork had already been laid. In the library, next to a book I was looking for, I saw Raimo Tuomela's *The Importance of Us* (1995). This was my "light bulb" moment. The relationships between institutions, group belief, and group action became much clearer.

This book applies some concepts from Tuomela's theory to my questions about institutions and knowledge. My primary concern is to elaborate some of the institutional mechanisms through which institutions shape the generation of new knowledge, particularly scientific knowledge. I have also worked to develop a conversation with constructivists in science and technology studies, whose influence on the text will be obvious (particularly in chapter 2). Nevertheless, the manuscript fits squarely within the framework of new institutionalism (Young 1994). My hope is that my colleagues will accept this manuscript as an extension of institutional theory (including neoliberal institutionalism), and that some constructivist colleagues will see points of intersection with their own work.

I

Theory

1

Introduction

During the past century, an important contributor to climate change and species extinction has been the collective, if unintentional, activity of human beings (Houghton et al. 1995; NRC 1995). In this respect, previous episodes of global warming, including that which ended the last ice age 11,000 years ago, differ from today's (Turekian 1996, 82). Likewise, previous mass extinctions, such as the one that destroyed most forms of life in the oceans some 225 million years ago, differ from today's massive loss of species (Gould 1977, 134). Today, human actions are important drivers of global environmental change. Given that some of the most pressing global environmental problems have an anthropogenic component, it seems reasonable to assume that what humans *think* about environmental phenomena will shape their *actions* toward them.

But how do people—groups of scientists, diplomats, fishers, or environmental activists—come to embrace certain understandings and not others? The present manuscript focuses on certain *institutional mechanisms* through which humans generate knowledge and accept beliefs about phenomena like the earth's ecosystems. Beliefs influence the ways in which agents form interests and select actions.[1] Of course, better knowledge of what ecosystems are, how they function, and their potential value to humans will by itself not eliminate anthropogenic threats to them (see, e.g., NRC 1995). But better knowledge about these phenomena will help humans to recognize their collective interest in protecting them.

This book illustrates how global and other institutions played an important role in generating knowledge about the marine environment at the Scripps Institute of Oceanography (SIO) and the Inter-American Tropical Tuna Commission (IATTC). In particular, it demonstrates how a specific set of institutional mechanisms—the positional fix, the statutory fix, and the committee fix—provided the means through which changes in institutions shaped the generation and use of knowledge at the Scripps Institution and the IATTC. These mechanisms are also shown to provide the means through which groups and organizations like the IATTC form beliefs and ground regulatory actions. A close examination of the history of research at SIO and the IATTC reveals the operation of these institutional mechanisms and allows us to distinguish them from the mechanisms posited by neorealist and interest-based theories in international relations.

1.1 Intended Contribution

In recent years, students of environmental governance have focused on how new scientific discoveries catalyze political action. For example, the discovery of the "ozone hole" in 1985 led to effective global regulation under the auspices of the 1987 Montreal Protocol and its subsequent amendments (Haas 1992; Litfin 1994; Benedick 1998). Scientific consensus about anthropogenic pollutants in the Mediterranean Sea, disseminated through an international epistemic community, accounts for the successful conclusion of the Mediterranean Action Plan (Haas 1990). In addition, new knowledge, disseminated by epistemic communities and bolstered by international institutions, stimulated international action to address climate change and acid rain (Social Learning Group 2001). Furthermore, active epistemic communities not only improve cooperation but also tend to improve compliance with international environmental regulations (Haas 1998).

In international relations much less attention has focused on the ways in which institutions, once formed, shape the evolution of knowledge, including scientific knowledge. Simply stated, the existing literature explores how scientific knowledge shapes international cooperation and compliance or the formation and functioning of institutions. But the question should also be turned around. How do global political and economic institutions, or issue-specific treaties like the Convention for the Creation of an Inter-American Tropical Tuna Commission, shape knowledge generation? It is this question that the present study seeks to address. The question is important because new knowledge, such as awareness of previously unknown threats, can reshape people's perceptions of their interests, and their actions.

Institutional mechanisms can be seen as one part of a complex web of social practices that shape the generation and use of scientific knowledge. Complex systems tend to exhibit nonlinear behavior. Incremental changes in some parts may yield sharp, strong effects in the whole (Holland 1995). Viewed in this light, institutions can be seen as "amplifiers" of certain ideas, that is, selecting some and magnifying their effects (Haas and McCabe 2001). Institutional theory cannot provide precise predictions about the content of knowledge yet to be generated. At the same time, it is possible to highlight certain mechanisms through which institutional changes shape the generation of and use of knowledge. This is most important in situations characterized by uncertainty. In such situations, people must commit to a particular belief (although they cannot be sure it is true) before they can recognize what their interests are or what their actions should be.

This book illustrates how groups use a specific set of institutional mechanisms to form beliefs. The mechanisms are the positional fix, the statutory fix, and the committee fix. Through these mechanisms, institutions shape the generation of new knowledge, often over long periods of time. Once formed, accepted beliefs repair uncertainty (at least temporarily), illuminate interests, and point to preferred actions. These mechanisms help us to explain the institutional dimension of knowledge generation and, in particular, how global institutions influenced knowledge production at the Scripps Institution and the IATTC. The mechanisms also allow us to explain the sense in which groups can be said to have beliefs and act in more than a metaphorical sense.

1.2 The Conceptual Domain

The international relations literature embraces Max Weber's (1913) insight: "Not ideas, but material and ideal interests, directly govern men's conduct. Yet very frequently the 'world images' that have been created by ideas have, like switchmen, determined the tracks along which action has been pushed by the dynamics of interest" (280). The question is not whether ideas *or* interests matter: both do. Assuming Weber points in the right direction, numerous questions remain. To take just two: If global political institutions shape the generation and use of new knowledge, how (by what mechanisms) does this happen? When decisions must be made in a context of uncertainty, can we identify institutional mechanisms that people use to grasp onto particular beliefs, at least temporarily, to decide?

The institutional literature points to a wide variety of mechanisms that may potentially shape the generation and use of knowledge. In this section I outline the broad domain within which the present work is situated.

An Institutional Approach to Knowledge

Institutions are "systems of rules, decision-making procedures, and programs that give rise to social practices, assign roles to participants in these practices, and guide interactions among the occupants of the relevant roles. Unlike organizations, which are material entities that typically figure as actors in social practices, institutions may be thought of as rules of the game that determine the character of these practices" (Young, ed. 1999; Young 1994).

Institutionalism, as defended in the present book, encompasses both the new institutionalism in the social sciences (e.g., Young 1994) and more sociological approaches (e.g., March and Olsen 1998). It draws from, and is consistent with, neoliberal institutionalism (NLI) (see Keohane and Nye 1989; Haas, Keohane, and Levy 1993) to the extent that there is agreement on certain foundational assumptions.[2] It also draws from, and is consistent with, weaker constructivisms, but not with stronger ones.[3] Institutionalism in a wide sense embraces three strands of theory: new institutionalism, NLI, and weak constructivism.[4]

A Brief Typology of Institutional Mechanisms

The institutional literature contains a wide variety of mechanisms that may potentially shape the generation and use of knowledge. By surveying some of these mechanisms, it is possible to outline a broad domain of present and future research on the knowledge-generating function of institutions. The following sections are not a summary of research results but rather a conceptual domain of which subsequent chapters develop one part.

Incentive Mechanisms To many institutionalists, the primary mechanisms of concern are incentives. Societies create property rights, which can be seen as a bundle of rights and responsibilities assigned to a property holder (North 1981; McCay and Acheson 1987; Ostrom 1990). These rights, in turn, create economic incentives that may encourage (or discourage) certain actions, like the generation of new technologies. For example, in the 1980s the prospect of global regulation of ozone-depleting chemicals (ODCs) reshaped incentives to chemical firms, essentially forcing them to increase investment in substitutes for ODCs (Makhijani and Gurney 1995). Framers of the Kyoto Protocol on climate change attempted to create market-based mechanisms, like tradable permits, to provide firms with incentives to develop lower-emission technologies (Young, ed. 1999; Victor 2001). In these examples, institutions shape knowledge by creating incentives for individuals or for firms.

Collective Action Mechanisms Sometimes the incentive structure is such that individuals do not produce the kinds of knowledge society needs. In general, failures of collective action refer to situations in which incentives to individuals lead to outcomes that are detrimental to society as a whole (Axelrod and Keohane 1986). Failures of collective *epistemic* action refer to situations in which incentives to individuals fail to generate the types of information that members of society collectively require. An important segment of the regimes literature explores how international organizations can remedy failures of collective epistemic action.

In such situations, international regimes can generate the kinds of information necessary to improve cooperation (Keohane 1989; Krasner 1983). International organizations may facilitate communication among stakeholders, raising concern about a particular problem (Haas, Keohane, and Levy 1993). For example, the United Nations Environment Programme has disseminated information about problems like stratospheric ozone depletion, thereby catalyzing international cooperation (Haas 1992; Litfin 1994). Organizations may design rules to improve transparency, or information about how well members are complying with rules (Mitchell 1994; 1998; Chayes and Chayes 1995; Hønneland 2000). They may also amplify the effect of epistemic networks, for instance, by developing procedures for integrating scientific advice into international policymaking (Haas and McCabe 2001; Andresen et al. 2000). In the past decade, a substantial part of the regimes literature has explored mechanisms to remedy failures of collective epistemic action.

Other Mechanisms Quite apart from incentives to individuals, it is possible to identify mechanisms through which institutions shape what people believe. Consider international judicial institutions. In the post–Cold War era, international judiciary bodies have proliferated (Keohane, Moravcsik, and Slaughter 2000). International dispute settlement panels establish rules and procedures through which a panel is to arrive at a decision. The specific rules under which these panels operate tend to vary widely (Keohane, Moravcsik, and Slaughter 2000, 459–468). However, once a panel has reached a decision, its ruling forms a precedent that may shape subsequent beliefs about the matter in question. This may happen formally, through legal precedent, or informally, by shaping public opinion. International judicial institutions may be an important, yet so far unexplored, mechanism shaping the formation of beliefs in international affairs.

Institutions may also generate beliefs about "social facts."[5] A social fact is a phenomenon, like trust or legitimacy, that would not exist in the absence of human society. If humans disappeared, so would the phenomenon. An example of a social fact is the use of paper as money (Searle 1995; Ruggie 1998). For paper to function as money, individuals in society must believe that it does. In addition, individuals must believe that other people will accept paper as money. Institutional rules, for example, that specially printed paper counts as money in a particular context, help to generate a collective belief in society that paper is money. Further, collective acceptance entails a reflexive element. An individual accepts paper as money, in part, because she believes that others in society will accept it ("I believe if you believe"). A second area for research concerns the mechanisms through which institutions create and sustain social facts.

Another type of institutional mechanism is the scientific mechanism. International organizations typically design standard procedures through which groups of scientists or experts reach decisions to advise policy-making. For example, the parties to the Framework Convention on Climate Change designed procedures through which scientists, organized into working groups, could arrive at a set of common beliefs with respect to the science of climate change (Andresen et al. 2000; Social Learning Group 2001). The end result of these standard procedures is one or more beliefs that the group (e.g., Working Group I of the Intergovernmental

Panel on Climate Change) accepts as a whole. A third broad area for future research concerns these standard procedures, or institutional mechanisms, through which groups of scientists accept beliefs.

The mechanisms identified here (and summarized in table 1.1) are ideal types. In the real world, they often appear together, nested one within the other. For example, a fishery regime might design an observer scheme to improve transparency (Mitchell 1998). Nested within the regime may be a dispute settlement panel to adjudicate alleged violations. Real-world institutions are complex, but identifying ideal types points to discrete causal mechanisms and possible directions for new research.

1.3 Specific Mechanisms of Concern

Section 1.2 described different types of institutional mechanisms that could, in principle, shape knowledge generation. The domain is large, and table 1.1 only begins to scratch the surface. In the pages that follow, three mechanisms will be put under a magnifying glass.

In general, the problem is to explain how people, at the group or social scale, generate knowledge and form beliefs. This is important because, in emergency situations, people must form some beliefs about situations in which they find themselves before they can decide what actions are in their interest. Often groups form beliefs routinely, for instance, when central bankers must decide what the rate of inflation is before setting interest rates, or when environmental managers must decide what the air quality is in a particular region before recommending regulation. Groups use institutional mechanisms to generate knowledge and fix beliefs. The term *fix* is used to emphasize that the mechanisms help *to establish common understandings* (at least temporarily), *to repair uncertainty,* and *to direct inquiry* (by fixing its direction).[6] Three mechanisms—the positional fix, the statutory fix, and the committee fix—will be examined in detail.

Before specifying these mechanisms, however, it will be useful to clarify briefly the manner in which the term *knowledge* is used in this book. For the purpose of the present study, *knowledge* refers to beliefs people accept in order to facilitate decision making. *Belief* refers to a statement that one or more people accept as true (Tuomela 2000b). Statements may be in verbal form, for example, disseminated at scientific

Table 1.1
Institutional Mechanisms for Knowledge Generation

Mechanism	Examples
Individual Incentive. This type of mechanism shapes the behavior of individuals and organizations. Individuals consider the incentives (payoffs) when choosing among alternative actions. Higher payoffs tend to encourage particular actions.	Institutions shape incentive structures within which individuals act (North 1981), e.g., rules can be designed to encourage or "force" the generation of new technologies (Benedick 1998; Makhijani and Gurney 1995). Ownership rights may be assigned in different ways (e.g., open access vs. private rights); changes in institutions of ownership shape society's ability to generate new technologies (Lessig 1999).
Collective Action. This type of mechanism develops rules and social practices to remedy "Prisoner's Dilemma" failures of collective action, where the structure of individual incentives creates an outcome that is collectively suboptimal (Axelrod and Keohane 1986). Typically this involves the creation of formal or informal regimes (Krasner, ed. 1983; Ostrom 1990; Young 1994).	Circulate knowledge about problems like stratospheric ozone depletion, or information to improve transparency about how well members are complying with rules (Mitchell 1998; Chayes and Chayes 1995).
Judicial. This type of mechanism typically involves some group or social decision regarding factual and normative disputes. Not primarily reflexive, in the sense that evidence is presented in support of (or against) candidate beliefs.	Group belief or acceptance, when institutionalized (e.g., decisions or rulings of war crimes tribunals or international criminal courts). Issues are typically related to violations of international treaties or laws.
Reflexive. This type of mechanism is concerned with the establishment of social facts such as trust and legitimacy ("I believe if you believe").	Use of specially printed paper as money in specified contexts.
Scientific. This type of mechanism comprises standardized procedures for accepting statements by scientists in aggregate. Statements are judged by standards set by the scientific community, such as conformity to logic or to observed phenomena.	A special kind of collective acceptance: peer review. A special kind of group belief: reports of scientific groups or committees, such as the Report of Working Group I of the Intergovernmental Panel on Climate Change.

meetings, in working papers, annual reports of scientific commissions, or peer-reviewed journals. They may also be in symbolic form, displayed in tables, graphs, charts, and maps. The term *knowledge* as used in this book does *not* refer to "justified true beliefs," the sense in which many philosophers use it.[7] For a more precise specification of *knowledge*, see chapter 2, section 2.3.

In this book, use of the term *knowledge* differs from its use in certain areas of philosophy and science and technology studies. For present purposes, questions related to justification and truth are bracketed. Also bracketed are questions related to the social construction of truth (see, e.g., Shapin 1994, 3–7; Barnes, Bloor, and Henry 1996). Using *knowledge* to refer to accepted belief opens a new path for research connected to the international relations ideas literature (Odell 1982; Haas 1990; 1992).

Given this understanding of knowledge, I turn now to specifying the mechanisms through which groups generate knowledge and fix beliefs. First is the *positional fix*. When using the positional fix, a person uses his or her social role, with attached rights and rules, as a guide when framing research or when selecting beliefs. Two examples illustrate the point. First, consider a scientist who serves on a committee to investigate the causes of a particular cancer. She may refer to the responsibilities attached to her role as committee member when framing her research. Even though she could in principle investigate anything, she tailors her research to generating knowledge consistent with the role and its responsibilities. Second, consider a scientist holding a position as chair of an environmental assessment committee. Although he cannot be 100 percent certain about a particular finding, as chair of the committee and for purposes of committee business he will tend to select beliefs consistent with prior published reports of the group. The positional fix refers to use of role or position to establish a belief (at least temporarily) or to fix the direction of research on the basis of one's role or position. As I show later, it is through the positional fix that the directors of the Scripps Institution and the IATTC influenced the direction of knowledge generated at their organizations. Changes in the role of the Director and associated responsibilities were reflected in the nature of the knowledge generated by the organization.

The second mechanism is the *statutory fix*. A statutory fix involves establishing one or more beliefs (at least temporarily) based on ideas embedded in formal or informal rules. John Ruggie (1983) referred to the economic framework embedded in the General Agreement on Tariffs and

Trade (now the World Trade Organization Agreements), a foundation of the world's open trading system, as "embedded liberalism" (188–189). Goldstein and Keohane (1993, 13) found that when they are embedded in rule systems, ideas shape policy outcomes. Embedded ideas can be used as a guide to research or the acceptance of particular beliefs. For example, a group of scientists could use a number of different models in their research (ecosystems models, maximum sustainable yield models). If their organization's foundational documents embed a particular framework, it will point researchers in one direction rather than another. The framework does not determine the content of research, but it privileges one approach at the expense of others. In the absence of the statute, scientists may well have selected a different framework. When they use ideas embedded in rules to guide research or to select beliefs, scientists and other experts employ a statutory fix.[8] This mechanism played an important role in determining the nature of the knowledge generated at the IATTC (see chapter 6). Changes in the ideas embedded in the treaty rules were reflected in the types of knowledge generated and accepted by the organization.

The third mechanism is the *committee fix.* A group of scientists or other experts may meet regularly as a group or committee in order to reach consensus (at least temporarily) on certain matters of fact. For example, in 1988, the Intergovernmental Panel on Climate Change was created to (among other things) assess available information on climate change (Houghton et al. 1995, foreword). The U.S. Federal Reserve regularly issues reports assessing the state of the U.S. economy. If such a group meets on a regular basis, formally or informally, it typically establishes regularized practices through which it accepts (as a group) some set of statements on some matter of uncertainty or some contested fact or set of facts. These rules and practices constitute the committee fix, through which groups form beliefs. As discussed in chapter 6, this mechanism was employed by the IATTC on a regular basis to establish group belief about the state of fish stocks and bycatch, and to recommend regulatory action.

These mechanisms are further discussed in chapter 2, section 2.2.

1.4 Outline of the Book

Section 1.5 seeks to make explicit certain assumptions that underpin the institutional approach adopted in this book. This is necessary because some of the assumptions, such as those pertaining to social ontology and social epistemology, differ from standard regime theory. It is also necessary to explain the sense in which institutionalism and poststructuralism are incommensurable (making empirical comparisons unhelpful, if not impossible). Readers primarily interested in the main argument may wish to skip directly to chapter 2.

Chapter 2 introduces political, economic, and epistemic institutions as complex systems. It specifies the mechanisms through which these institutions shape the generation and use of knowledge, and clarifies how the term *knowledge* is used. It then explains the research method: structured, focused case studies.

Chapters 3–5 present a focused case history of the Scripps Institution of Oceanography from about 1900 until about 1970. The case was selected because it allows for variation in global political institutions, such as the emergence of the United States as a world power (and regional hegemon) after 1890 and the emergence of the United States as a global hegemon after World War II. Successive Directors of the Scripps Institution used the role of Director (which overlapped with political roles like member of the National Research Council) to frame and reframe research over time. Agents translated changes in global political institutions into changes in knowledge, using the positional fix (as will be argued, referring to the U.S. role in the world, mediated by committees including the National Research Council). A traditional approach to international relations—neorealism—cannot account for this change.

Chapter 6 provides a structured, focused case history of the IATTC. The IATTC used two mechanisms—the statutory fix and the committee fix—to form beliefs about the status of the stocks. As these beliefs changed, so too did the organization's regulatory actions. As the IATTC became aware of increased anthropogenic threat to marine life (yellowfin and dolphins), its regulatory actions became stricter. Another influential view—the interest group approach—cannot explain these changes in knowledge or action.[9]

1.5 The Institutional Perspective

Exploring the knowledge-generating function of institutions requires charting new conceptual terrain.

To orient the analysis it is useful to start with a distinction between collective action and social practice approaches within the new institutionalism (Young 2001). As the analysis moves to less familiar ground, it is necessary to make certain foundational assumptions explicit. Accordingly, this section positions the book with respect to collective action and social practice approaches, clarifies certain ontological and epistemological assumptions, and locates the analysis within the new institutional tradition, broadly defined.

Foundational Assumptions

Young (2001, 9) divides the new institutionalism into two competing families: collective action and social practice.[10] There are three main differences. First, collective action models assume identities are fixed for purposes of analysis. Social practice models focus more on the social construction of identities, for instance, the ways in which regimes shape the identities of participants. Second, collective action models assume that behavior stems from utilitarian calculations ("logic of consequences"), whereas social practice models emphasize the importance of social norms ("the logic of appropriateness") (March and Olsen 1998). Third, collective action models tend to abstract from contextual details, facilitating formalization. Social practice models emphasize the importance of particular contexts, favoring thicker description and qualitative analysis. While there is considerable variation among specific models within these families, Young argues, institutionalists generally tend to identify themselves with one or the other.[11]

The present work does not fit neatly into either family but incorporates elements of both. First, it is concerned less with identity than with social roles.[12] As institutional artifacts, roles tend to be relatively stable. At the same time, roles are human creations ("socially constructed") and are subject to change over time. Rather than assuming roles are fixed (as collective action models might) or asking how roles are constituted in particular contexts (as social practice models might), the present study investigates the question, What is the effect of changes in social roles on the generation of knowledge? That is, how do social roles mediate the generation of knowledge?

Second, when a person acts as a group member, that is, within a social role, both consequences ("payoffs from actions") and appropriateness ("normative correctness") matter. This assumption appears throughout the case studies. For example, when the founders of the Scripps Institution created the role of Director, both consequences (research results) and appropriateness (scientists *should* conduct research) mattered. The Institution's early benefactors, in turn, were concerned not only with the discovery of new forms of life in the ocean but also with their conviction that the Director *should* generate research for the benefit of society. In short, it is assumed that logics of consequences and appropriateness were built into the Director's role.

Third, context matters. Following Turner (1985) and Tuomela (1995; 2000a), I assume that in some contexts a person will act as an individual, and in other contexts she will act as a group member. Of course, formal models necessarily abstract away from contextual factors and cannot capture context as well as well-written case studies can.[13] Nevertheless, it is possible to capture salient elements of context in formal models and to test them with laboratory-type experiments (see, e.g., Turner 1985; Tuomela 2000a). In short, the present analysis does not fit neatly into either family, collective action or social practice. It incorporates elements of both.

To differentiate the present analysis from other approaches, it is necessary to focus on specific models within these families. The interest group model, equivalent to neoclassical economic analysis applied to politics (Olson 1965; Moravcsik 1997), falls within the collective action approach. So does neorealism (Waltz 1979; Mearsheimer 1994).[14] Poststructuralism, which applies insights from the later Foucault, can be placed within the social practice category (Walker 1993; Litfin 1994). The following sections seek to clarify the differences between institutionalism and these alternative approaches, by exposing certain differences in foundational assumptions.

Assumption 1: Social institutions exist. Social institutions exist and tend to generate distinct kinds of social order. *Social order* refers to patterns or regularities that emerge in human behavior, and which often persist over time (Bull 1977; Holland 1995). In international affairs, where governmental arrangements do not exist, people develop and use institutions to provide a measure of order or governance (Young 1994). Institutions are rules and social practices that account for particular kinds of order.

For example, a community may set limits on who may fish in a certain area, thereby designating certain people as fishers, with rights to fish. Attached to the role of fisher is the responsibility to observe restrictions the community may impose, such as the kinds of gear that can be used or open and closed seasons. In situations in which fishing grounds are in international waters or span national boundaries, an international organization like the Form Fisheries Agency or the Inter-American Tropical Tuna Commission may be required to recommend regulations. Whether or not the particulars are written down, active institutions entail assignment of roles, to which rules and rights are attached. Institutions thereby generate patterns in human behavior.

By contrast, the interest group approach assumes either that social institutions do not exist or that they are essentially fixed over long periods of time. Within a given institutional structure such as private property rights, questions focus on how individuals maximize utility subject to constraints. Individuals respond to (exogenously fixed) incentives when choosing among alternative actions. Agents choose the actions that will generate the highest payoffs (Hayek 1948; Olson 1965; North 1981). To return to the example of fishers, the interest group approach assumes that, in the absence of some form of coercion, individuals will overfish to the point where the stock collapses (Hardin 1968; critique in McCay and Acheson 1987).

As applied to knowledge, the interest group approach assumes agents generate the kinds of information that enables them to maximize utility. Friedrich Hayek, for example, assumed that in competitive markets, individual buyers and sellers generate all the information they need to conclude transactions. In aggregate, across society, individual agents generate the socially optimal amount of information "as if" guided by an invisible hand (Hayek 1948). According to this approach, if agents do not produce a certain kind of information—for example, a catalog of species in a certain marine area—then it was not in their interest to do so.

Poststructural approaches provide a second contrast. Consistent with the later Foucault, poststructural approaches assume that institutions are epiphenomenal to discourse (Dreyfus and Rabinow 1982, 77). In other words, as discourses shift, so will institutional configurations. For example, Litfin (1994) analyzed the construction of discourse associated with the discovery of the "ozone hole," and how a discursive shift associated with this discovery made the negotiation of the 1987 Montreal Protocol possible.

The chasm separating institutionalism and poststructuralism cuts deeper.[15] Rather than seeking to discover pattern or order (as this book does), poststructuralists tend to emphasize discontinuity, contingency, and process. What institutionalists see as institutions and knowledge are to poststructuralists fluid power/knowledges, mutually embedded knowledge/orders, or discourses. Turning attention away from structures (like institutions), for example, Litfin writes, "Structures, constituted by identities and interests, cannot exist apart from process" (Litfin 1994, 3). To some, the goal of analysis itself is to destablilize existing conceptualizations (Walker 1993, 25).

Because they differ in this core assumption, poststructural approaches are incommensurable with institutionalism. Differences in foundational assumptions (including, but not limited to, "institutions exist") make it impossible to translate observations from one to the other. This does not mean poststructuralist approaches are wrong. But they are sufficiently different to render comparative analysis impossible.[16]

Assumption 2: International structure comprises socially constructed rules and practices, not simply material capabilities. A second assumption is that political order can usefully be understood as an institutional system. Following a path opened by Bull (1977), "hegemon" or "great power" can be seen as a social role that an agent takes on, with certain rights and responsibilities attached. Although preponderant military capabilities are necessary for a state to acquire a role as hegemon, they are not sufficient: the state must accept the role (see section 2.1). Within this approach, *international structure* refers to a complex set of overlapping political, economic, epistemic, and other institutions, in which agents relate to each other through relatively stable sets of roles, rules, norms, and social practices.

Assumption 2 is consistent with the work of some constructivists. For example, Wendt (1992) argues that the international political structure assumed by neorealists is composed of social practices, not material capabilities. At stake is a social rather than a strictly material ontology.

The second assumption is inconsistent with neorealism. Waltz (1979), Mearsheimer (1994), and others in this tradition assume that international structure is primarily composed of material capabilities, or physical measures of its ability to successfully wage war.[17] These include stockpiles of weapons, economic productivity, population, and relevant natural resources. If a single state dominates, the structure is unipolar; two states, bipolar; and more than two states, multipolar. According to this model, as in imperfect markets where oligopolists or monopolists set prices, great powers determine broad patterns in international politics (see, e.g., Waltz 1979, 91–101).

Finally, although assumption 2 differs from that of neoliberal institutionalism and regime theory, it can be seen as a friendly amendment. While material capabilities matter, international political structure is, at root, institutional. Within a framework of meta-institutions like sovereignty, property rights, and scientific disciplines, more specific regimes are nested. This is broadly consistent with the new institutionalism in the social sciences (Young 1994; Ostrom 1990; McCay and Acheson 1987).

This assumption stands out most clearly in chapters 3–5. That the Scripps Institution's research program was, over time, shaped by political factors is hardly news (Raitt and Moulton 1967; Mukerji 1989). However, the existing literature does not explain very clearly *how* the state shaped the research program. Increased funding is one obvious factor. But an institutional approach enables the analyst to identify specific social mechanisms through which political changes reshaped research practices.

Assumption 3: People can act as individuals or as group members. A person can act in a personal capacity as well as in a social capacity (Tuomela 1995; Turner 1985; Turner et al. 1994). When a person acts "as Chairman," or qua group member, she acts for the group. The same person, in a different context, may act qua an individual (in a personal capacity). When acting as a group member, a person's actions are primarily shaped by her social role. When acting as an individual, the same person's actions are more strongly shaped by self-interest.

Evidence from social psychology suggests that, depending on context, a person will activate either a social or an individual identity (see Turner 1985; Turner et al. 1994; Haslam et al. 1996). In support of this view, Turner (1985) reports evidence from a series of experiments with Prisoner's Dilemma games.[18] When both players perceived themselves to be part of the same group, their choices were twice as cooperative compared to players that perceived themselves to be members of different groups (Turner 1985, 88). In other words, when acting qua group members, players chose to cooperate twice as often compared to players acting qua individuals.

Assumption 3a: Groups as well as individuals can act. Groups as well as individuals can act. A group can be said to act when one of its operative members, acting qua group member, acts. An operative member is an individual empowered to act for the group (Tuomela 1995). A Prime Minister, for example, is an operative member of a state.

This assumption stands in contrast to methodological individualism, which appears in stronger form in the interest group approach. In its stronger form, methodological individualism assumes that *only* individuals can act (Hayek 1948; Yarbrough and Yarbrough 1990, 242). States are modeled *as if* they were rational individuals, and real-world diplomats are individuals who act to advance the interest of the state.

Assumption 3a underpins discussion throughout chapters 3–5. For example, in chapter 3, the Scripps Institution's research program develops as a holistic enterprise, not merely as the sum of individual lines of work. The Director (in the early days, William E. Ritter) created a role for himself as the person who planned and directed this organized group. It is assumed that the Scripps Institution (as an organization) can conduct research. Ritter, acting qua Director, can act for the group as a whole.

Assumption 3b: Groups as well as individuals can hold beliefs. Groups can hold beliefs. This assumption is adapted from the concept of positional group belief (Tuomela 1995). For the purposes of group action, when acting in group mode, members may accept one or more beliefs. This happens, for example, when a regulatory organization must establish some belief about the state of the environment before recommending new rules. Group beliefs are often expressed as the report of a committee, for instance, the report of Working Group I of the Intergovernmental Panel on Climate Change, or the Organization for Economic Cooperation and Development's report on future trends in members' economies. Assumption 3b reflects a social as opposed to a strictly individual epistemology.

It therefore contrasts sharply with strong methodological individualism as seen in the interest group approach. In its strong form, individualism insists that individuals are the *only* agents that hold beliefs. Self-interest leads individuals to form particular beliefs, for example, about market price.

According to this view, it may appear that individuals in a group have formed a social belief, since individuals may all believe the same thing. However, through the "invisible hand" of competition, self-interest directs each individual, pursuing his own interest, to form his own belief. If we assume—as the interest group approach does—that the world is transparently knowable, it is not surprising that self-interested individuals simultaneously arrive at the same belief (Hayek 1948, 33–56).

This invisible hand explanation underpins the economists' assumption of perfect information in perfectly competitive markets. To strong individualists, social knowledge cannot exist except as an aggregation of the beliefs of individuals. Gaps in knowledge are always reducible to a failure of collective epistemic action, a situation in which individuals, pursuing their self-interest, fail to produce a socially optimal outcome.

Social epistemology is consistent with the assumption of weak individualism. All that is required is recognition that under certain circumstances individuals accept beliefs qua group members. For example, the Catholic Church believes that miracles happen. As a Catholic (qua a member of the Church), John believes that miracles happen. Or: the Communist party of state X believes that capitalist countries will soon perish (though none of its members believes so). As a member of the Communist party of X, John believes that capitalist countries will perish (as an individual, he does not) (Tuomela 1995; 2002). In these examples, beliefs are social in the sense that the organization plays a role in forming them. The group does not add up the beliefs of individual members to arrive at a belief. Nothing in this formulation denies that in some situations group beliefs can be formed by canvassing beliefs of members. The point is that other methods of arriving at social beliefs also exist.

In the case studies, the notion of group belief appears most clearly in chapter 6. I assume that it is possible for the IATTC to express beliefs as an organization. Key observations in chapter 6 are reports from the IATTC's scientific staff, including tables, graphs, and other statements from its Annual Report.

Assumption 4: Causal inquiry is valid in the social sciences. It is possible to frame causal hypotheses about social phenomena and to evaluate them using the methods of scientific inquiry. The interrelationships between institutions and social knowledge appear to be varied and complex, but this complexity itself does not negate the possibility of causal analysis (King, Keohane, and Verba 1994, 10–12, 42–43).

Causal here means that changes in one specifiable phenomenon tend to stimulate changes in another, through mechanisms we can identify. It does not mean that these mechanisms fully determine the content of new knowledge. Its evolution involves pattern and order as well as spontaneity and chance. Nor does it mean that "mechanisms" are immutable facets of human nature or of natural law. The mechanisms of concern have been built by people for the purpose of facilitating reasoned (and principled) social action.

In outlining the research agenda for institutionalism in the social sciences, Oran Young (1994) writes, "Above all, we need to investigate the behavioral mechanisms through which institutions produce their effects. . . . Unless we understand the causal connections involved in these impacts, arguments regarding the significance of institutions as determinants of collective outcomes cannot progress beyond correlational accounts" (8).

Many constructivists, on the other hand, seem to avoid causal analysis (e.g., Litfin 1994, 7). In general, constructivists in international relations have focused on constitutive questions, including the processes through which identity, authority, and meaning are formed (and reformed). However, it is also valid and important to identify causal mechanisms through which institutions shape knowledge.

Assumption 5: A world external to the human mind exists; human agency is required to grasp it as knowledge. Assumption 5 is that there is a world out there, and that that world is at least partially knowable. In other words, phenomena external to the human mind (like fish or dolphins) exist, independent of human beliefs about them. Scientific statements can reflect phenomena in the world, albeit with uncertainty. At the same time, the generation and acceptance of beliefs requires human agency. Furthermore, the processes through which beliefs are taken up to inform action are subject to political contestation (Jasanoff 1997; Social Learning Group 2001).

Table 1.2
Ontological and Epistemological Assumptions

Assumption	Contrast
1. Social structures (e.g., institutions) exist and can be relatively stable over time. Institutions, among other things, shape incentives (e.g., through particular configurations of property rights) (North 1981; Ostrom 1990).	According to the interest group approach (Hayek 1948; Olson 1965), incentive structures are essentially fixed. Questions turn on how individuals maximize utility subject to constraints. According to *strong constructivists* and *postmodernists* (e.g., the later Foucault), discourses are continually remade and reinterpreted, and institutions are continually in flux. Questions turn on the construction of (fluid and contingent) identities, authorities, and discourses, not on the effects of relatively stable institutional configurations.
2. The international political and economic orders are institutional (socially constructed).	*Neorealism* (Waltz 1979; Mearsheimer 1994) assumes that the international political order is primarily composed of material phenomena, e.g., material capabilities to wage war.
3a. Social ontology: groups as well as individuals can act (Tuomela 1995; Turner et al. 1994). Consistent with *weak individualism.*	The *interest group approach* (Olson 1965; Moravcsik 1997) assumes only individuals can act. *Strong methodological individualism* models states and other groups "as if" they were rational individuals: only individuals can act (Yarbrough and Yarbrough 1990).
3b. Social epistemology: groups as well as individuals can hold beliefs. Consistent with *weak individualism.*	According to the *interest group approach,* only individuals can hold beliefs ("a belief is an idea in the mind of an individual"). *Strong methodological individualism* models states and other groups "as if" they were rational individuals: only individuals can accept beliefs.

Table 1.2 (continued)

Assumption	Contrast
4. Causal inquiry is valid as applied to human actions (King, Keohane, and Verba 1994).	*Strong constructivists* favor constitutive questions, emphasizing the constitution of identities, authorities, or meaning.
5. A world external to the human mind exists; human agency is needed to grasp it as knowledge. Phenomena exist independent of human beliefs about them. Scientific statements reflect phenomena with uncertainty. Generation and acceptance of beliefs require human agency.	According to *deep relativists,* science cannot inform the policy process because it is determined by political interests. According to *naive realists,* the authority and integrity of science depend in part on its autonomy from political influence.

Assumption 5 is inconsistent with stronger forms of realism in the philosophy of science. The realist tradition maintains that proper scientific inquiry is objective, divorced from political interests or ends (Popper 1972, 56). In this view, the authority or integrity of scientific research depends in part on its autonomy from political influence (Andresen et al. 2000). When taken to an extreme, this position is known as naive realism (Jasanoff 1997).

Assumption 5 is also inconsistent with radical relativism. Radical relativists maintain that scientific statements cannot inform the policy process because scientific research itself is saturated with political interests. Broadly speaking, socially mediated realism as applied in this book will not please stronger constructivists.[19] It does not assume that beliefs and interests, or knowledge and power, are inseparable, for instance, as power/knowledge.

This assumption informs much of the discussion in chapters 3–5. The cases are not meant to suggest, as a naive realist might, that political or economic influence corrupted what could have been a "pure" research program at the Scripps Institution or the IATTC. Nor are they meant to suggest, as a radical relativist might, that research findings are nothing but discourses that reflect power configurations.

Table 1.2 summarizes these assumptions and the contrasting views of them held by various approaches to international relations.

Institutional Framework

Of central concern to the institutional research agenda is governance. Institutions are sets of rules that give rise to social practices, create roles, and guide interactions among occupants of relevant roles (Young 1999; Young, ed. 1999; March and Olsen 1989, 160). A complex, overlapping set of social institutions provides a measure of governance in international affairs.

Institutions shape social practice as follows. When deciding how to act, a person evaluates the context. Does the context stimulate a person to act qua an individual or qua a group member? (Tuomela 1995; 2000a; Turner 1985). If she acts in group mode, what role is the person expected to fill, and what are the rights (or obligations) associated with this particular role in this particular context? In unfamiliar contexts, people tend to refer to what they know about social institutions (roles, with attached rights and obligations) for guidance. Therefore institutions generate order, or patterns, in social behavior (March and Olsen 1989; 1998; Turner et al. 1994; Turner 1985). Institutions also provide flexibility and guide people in new situations or in uncertain contexts.[20]

The basic contours of the new institutionalism are well known in international relations (see, e.g., Young 1994; Yarbrough and Yarbrough 1990; Keohane 1989) and will not be elaborated here. However, it is necessary to differentiate institutionalism from a standard interest group approach.

In its most extreme form, the interest group approach assumes that institutions do not (or need not) exist. It assumes that patterns in social action that appear to result from conscious human coordination actually result from simultaneous, independent decisions by self-interested individuals. Buyers and sellers generate enough information about market price (and other relevant parameters) to conclude transactions because each is propelled to do so by an "invisible hand" (Hayek 1948, 33–56). In competitive markets, those with inadequate knowledge are eliminated. In less extreme form, the interest group approach constitutes a "thin" institutionalism, studying cooperation among self-interested individuals (Krasner, ed. 1983; Oye 1986; Keohane 1989; Milner 1997). To state this more clearly, the interest group approach assumes an individualist ontology. In stronger form, it assumes a strong methodological individualism, that is, "only individuals can act" or "only individuals can hold beliefs" (Yarbrough and Yarbrough 1990, 236). By contrast, the present analysis assumes that groups like organizations can act and that they can accept beliefs. This occurs through the agency of individual acting qua group members. At the same time, it is useful and valid to study the actions and beliefs of individuals acting *qua* individuals. Compared to the interest group approach and to traditional regime theory, the present analysis is a more *social* institutionalism.

1.6 Summary

Beliefs matter in international politics. In situations characterized by uncertainty, the beliefs people accept shape their interests and actions. Whereas most of the international relations literature focuses on how ideas affect international cooperation and compliance (Haas 1990; 1992; 1998; Goldstein and Keohane 1993; Litfin 1994), this book turns the causal arrow the other way. It focuses on how institutions shape the generation and use of knowledge.

This chapter provided a rough map of the conceptual terrain by surveying the types of mechanisms through which institutions shape knowledge. It clarified five ontological and epistemological assumptions on which the analysis in this book is based. The assumptions underpinning institutionalism were compared with the interest group approach, neorealism, and poststructuralism. Subsequent chapters compare an institutional approach to potential alternatives: neorealism and the interest group approach. Poststructuralism, which is incommensurable with institutionalism, cannot be evaluated in comparative terms.

Chapter 2 clarifies the sense in which complex political, economic, and epistemic systems can be seen as institutional. It introduces the three institutional mechanisms that are the primary focus of this book: the positional fix, the statutory fix, and the committee fix.

2

Global Institutions and Social Knowledge

Uncertainty can corrode action. What if a person, exhausted and thirsty, finds a stream of water she thinks might be toxic? She must accept some belief about the water's safety—even if she can't be sure—before she decides whether to drink. Societies may face emergency situations like war, economic or ecological crises, in which they must form beliefs in the face of uncertainty, in order to act quickly. To form beliefs, societies create a variety of institutional mechanisms, which tend to shape the generation of knowledge over time. This chapter presents an institutional theory of knowledge formation. Its primary concern is with the institutional mechanisms through which groups of people (organizations, committees) fix beliefs in order to guide action.

Section 2.1 clarifies the sense in which global political, economic, and epistemic systems can be seen as institutional. It thereby differentiates the present analysis from standard regime theory in international relations. Section 2.2 specifies three mechanisms—the positional fix, the statutory fix, and the committee fix—through which global and other institutional factors shaped the generation and use of knowledge at the Scripps Institution and the IATTC. Section 2.3 explains how the term *knowledge* is used in the present work. Finally, section 2.4 details the research method: focused comparative case studies.

2.1 Global Institutions

A theory of political, economic, and epistemic systems as overlapping sets of institutional rules suggests that a complex set of social roles, rights, rules, and norms shapes human action, including the generation of new knowledge. This depiction of global institutions differs in important respects from neorealist theory (Waltz 1979; Mearsheimer 1994) and conventional regime theory (Krasner, ed. 1983; Krasner 1999; Keohane 1984). Readers who do not require clarification of the nature of these differences may wish to skip directly to section 2.2. The following paragraphs draw from institutionalism in the wide sense, extending it with concepts borrowed from philosophy, social psychology, and science and technology studies.

Global Political Institutions

According to institutionalism, the global system of states can be understood in terms of socially constructed rules and practices. Since the seventeenth century, states have regarded each other as the major actors in the global system. Each state, qua sovereign, has a right of self-determination that other states recognize, even if they do not always respect it in practice. A small subset of states may acquire roles as great powers, or hegemons.

Classics in the institutional tradition are usually read as requiring the existence of an international society. International society, or more precisely the expectations of the members of an international society, generate social roles and attached rights and rules. According to Bull (1977), international society refers to a society of states, where leaders of each state recognize each others' roles, rights, and responsibilities. A more recent interpretation holds that international society consists not merely of states but of other actors, including members of civil society, which is increasingly global (Young 1999).

One criticism of this approach has been that there does not exist a society at the global scale, in the sense that there is no recognized authority system or global culture. In what follows I show, based on the ideas of Tuomela (1995), that one does not need to posit the existence of a global society in order to preserve the key insights of institutionalism. What is required is that we distinguish between primary social roles, which require the existence of a society and a recognized authority system, and secondary social roles, which do not.

Primary Social Roles Just as individuals can acquire roles with respect to other individuals, groups can acquire roles with respect to other groups. According to Tuomela (1995), a social role in a primary sense presupposes the existence of a social group. In a wider, or secondary, sense groups can acquire roles with respect to each other. Roles in this secondary sense are generated by mutual expectations in the group that some division of tasks is necessary.

A social group (in a primary, or core, sense) is a collective that has acquired an authority system. Individuals in the group regard each other as members and thus acquire a "we-ness." In order to function or to act as a unit, the group must use the authority system to establish intentions. An intention is a kind of commitment, which can be expressed in the form "we will do X," that the group accepts (Tuomela 1995).

Social roles in the primary sense also presuppose the existence of rules and norms in the group. Rules are actions that the occupant of a role is expected to perform. Rules may be formal, that is, adopted constitutionally and written down, or they may be informal but activated through social practice. Social norms refer to actions that a group believes members should perform, in the sense that they *ought to* perform them. Some norms are common across human societies, like injunctions against murder, while some norms are group-specific.

Furthermore, while some rules and norms apply to everyone in a group, some are divided among group members. When rights and rules, or more informal tasks or responsibilities, are divided among group members, this division creates a social structure. It is the division or specialization that creates the structure. Structure can be overt, that is, created by formal or informal rules, or it can be covert, in the sense that it is created by social norms that are not written down.

A social role can be rule-related, that is, created by formal or informal rules. A social role can also be generated by social norms in the group. This happens when group members believe an agent should perform certain tasks, although there is no explicit rule specifying such an assignment.[1]

Secondary Social Roles It is also useful to think of sovereign agents as having one or more roles with respect to other agents in the global collective. The term *collective* is used to signify that at a global scale there exists a heterogeneous group of agents that do not form a social group in a primary or core sense.[2] In other words, agents in the global collective have not established a single authority system or government.

A state can acquire a role with respect to other groups in situations in which there is a division of responsibilities or tasks. This division generates a social structure in the collective. Such a division may occur through the threat of force, in which one or more agents provide deterrence or collective defense. Task division may also occur through formal agreement. This happens, for example, when states create regimes to deal with specific problems. An active network of such regimes provides some measure of governance in international affairs (Young 1999).

To be precise, division of tasks creates an *additional* role for particular sovereigns. All states occupy the role of sovereign (in a primary sense). When recognized by other states, for example, by membership in international organizations like the World Trade Organization or the United Nations, states acquire a secondary role as sovereign. In addition to this, some sovereigns acquire additional roles, for instance, as great power, as a consequence of task division. If the state is to acquire additional responsibilities, other agents must believe the state is able to perform the required actions. All states with international recognition occupy the role of sovereign; of this set, some acquire an additional role of great power.

Sovereign as a Primary Social Role The sovereign can be understood as having a primary role with respect to members of a core social group (i.e., the relation of an individual to a group). For example, a President or Prime Minister occupies a primary role with respect to citizens. In a primary sense, the sovereign can be understood as an agent empowered to act for a group. The operative agent of a state might, for example, be a President, a Prime Minister, an autocrat, or an oligarch. For the sovereign to be legitimate, group members must share a belief that the agent occupying the role of sovereign is subject to rules and has rights. Members must also share a belief that the agent is able to act in a way consistent with the rights and rules, and is disposed to do so. Moreover, group members must share a belief that the tasks and rights attached to the role of sovereign are consistent with the more general rules and norms active in the group. If these conditions do not hold, the sovereign is not legitimate.

Since social groups create roles, and attach rights and responsibilities to them, specific institutional configurations tend to vary. A considerable body of empirical work demonstrates that political institutions, like sovereignty, tend to change over time (Chayes and Chayes 1995; Kratochwil 1995; Ferguson and Mansbach 1996; Litfin 1998). Since the rights and rules attached to the sovereign role tend to change, it is not possible to provide an a priori specification of their content. In this sense, specific rights and rules attached to the sovereign role are an open class.

The core element of sovereignty in the primary sense is the delegation of authority, or the right to act on behalf of the group, to an operative agent. Therefore an agent (such as a state) can be understood as the occupant of an institutional role, sovereign.

The significance of this delegation of operative agency has been a classic problem for political theorists. Hobbes (1651) believed this transfer of authority was part of a historical social contract through which a society permanently surrenders absolute authority to a king or a legislature, by virtue of a need to end the "war of every one against every one" (100). To Vattel (1805), the transfer reflected a social contract adopted by rational individuals, which could (should) be canceled if the sovereign failed to fulfill the responsibilities assigned to the role. Institutionalism does not help specify the nature of the transfer of authority. However, it does specify that the transfer is inseparable from, and legitimized by, the group's assent.

Rights and Rules Attached to the Role of Sovereign In general, the rights and rules groups assign to sovereigns can be understood as generative structures. A small number of rules tend to guide a wide variety of behaviors. When agents find themselves in unfamiliar situations, these rules tend to guide their actions. Therefore, agents will tend to display patterned behavior in a wide variety of contexts. The relevant question for purposes of this study is the extent to which such actions include generation of knowledge and acceptance of beliefs.

Historically, a primary responsibility of the sovereign has been self-preservation. Vattel (1805, 62) wrote: "[The sovereign's primary] obligation is to preserve the duration of the political association on which it (the sovereign state) is founded." This includes defense against external invasion. To fulfill this obligation requires the sovereign to develop sufficient material capabilities. It also includes responsibilities toward group members, including enforcement of laws assigning ownership of property and upholding standards of justice. Hobbes (1651, 129) argued that the most important goal of the commonwealth was to maintain security. This basic assertion is essentially uncontested in international relations theory.

Self-preservation can be understood as a right or responsibility attached to the institutional role of sovereign. This means that the use of force is consistent with, and intrinsic to, the institution. Self-preservation has been a primary responsibility of sovereigns. As such, it is not only a rule or responsibility requiring the sovereign to defend a territory. The group also assigns to the sovereign the right to do so, in the sense that people consider actions taken to defend them to be legitimate. The society also attaches a normative injunction that the sovereign should defend the realm. Actions taken to defend them, including amassing of weapons, the use of force, coercion, and war, can be seen as the responsibility, the right, and the normative obligation a group assigns to the role of sovereign. Should groups routinely assign these kinds of responsibilities, rights, and obligations to agents *other than states,* this would constitute a change in sovereignty as an institution.

The Secondary Roles of Great Power and Hegemon According to Bull (1977), a small number of powerful states tend to maintain order in global politics by creating social roles for themselves. Roles are not descriptions of what states actually do because sometimes states do not act in ways consistent with them. Nor are roles prescriptions for what states should do. To interpret them this way is to beg certain questions about the moral status of a particular order (205–207).

Furthermore, the political role of a great power is multidimensional. An agent can take a number of actions to further the overarching goal of preserving a political order. These include, but are not limited to, seeking to avoid or to control crises among great powers; reciprocal agreement among great powers to respect each others' spheres of influence; and undertaking joint action with other great powers to preserve international order. These actions can be seen as rights or responsibilities assigned to the state's role.

Great-power status can be created by a threat of force. In this situation, the role is generated by the existence of an agent within the collective with a capability and intention to inflict harm and a mutual expectation in the collective that this is the case. Here the agent's (secondary) social role is a consequence of (1) the agent's, and the adversary's, brute capability to inflict force; (2) the existence of mutual expectations in the collective that this is the case. Structure emerges as a consequence of task division. The role (and the structure) will change if either (1) or (2) changes. This interpretation is generally consistent with Bull (1977), except the sovereign role is not rooted in an international society.

If great-power status may derive in part from a brute capability to inflict force, in what sense does institutionalism generate conclusions that are different from neorealism? First, to the extent that global politics entails military competition among states, this is a social phenomenon, only one of a set of possible ways global affairs might be ordered (Wendt 1992; 1999).

Second, and more directly relevant to this study, a state with preponderant military capabilities might not accept a role as a great power or hegemon. According to the neorealist model, international structure such as great-power status derives from brute capabilities alone. Drawing from institutionalists like Bull (1977), institutionalism maintains that the role of great power can be seen as a cluster of expectations, in the form of rights and responsibilities, that an agent can choose to accept or to reject. Therefore a state may have the brute capabilities of a great power, but not the role of a great power. Before World War I, and again in the interwar years, the United States possessed great-power capabilities but rejected the great-power role (Carr 1939; Wight in Toynbee 1952; Hillman in Toynbee 1952; Bull 1977, 212; Kennedy 1988, 328; Gaddis 1997, 6–7).

A great power may also occupy a role as hegemon. *Hegemony* refers to a situation in which a power dominates smaller powers in its sphere of influence but does not habitually use force to do so. Hegemony represents some middle range between a strong state assuming the right to use force in its relations with smaller powers, and a situation in which weaker powers willingly accept the leadership of the stronger. For example, historically, U.S. policy toward Latin American and Caribbean states tended toward the pole of habitual intervention and use of force to regulate the affairs of weaker states. On the other hand, U.S. relations with its NATO allies during the Cold War was closer to leadership (Bull 1977, 214–215). The role of hegemon is somewhere in between. The state may resort to force or the threat of force but does so only on occasion and prefers to exercise influence by other means. Hegemony, like great-power status, is a social role a state may acquire.

Global Economic Institutions

Like the role of sovereign, the role of property holder can be understood in two senses, primary and secondary. In a primary sense, a group assigns certain rights and responsibilities with respect to some thing to an agent. The agent to whom rights are assigned may be an individual, an organization like a state or firm, or a community. In a secondary sense, the property holder acquires rights and responsibilities with respect to agents outside the group. On a global scale, the 1883 Paris Convention, the 1886 Berne Convention, the 1982 UN Convention on the Law of the Sea, and the 1994 Trade Related Aspects of Intellectual Property Rights (TRIPS) agreement deal with rights in the secondary sense.

The agent to whom rights are assigned may vary. If rights are assigned to individuals or to firms, the institution is private property rights. Rights assigned to a state are state property rights. Rights assigned to a community are community or common property rights. If rights are assigned to no one, the institution is open access. This typology is standard in the institutional literature (e.g., see McCay and Acheson 1987, ch. 1; McCay 1996, 14; Ostrom 1990; Ostrom et al. 1999).

It differs from terms commonly used in economics. Economists tend to conflate social phenomena, like the assignment of rights, with qualities inherent in the phenomena themselves. For example, economists often assert that large-scale systems such as the atmosphere have intrinsic qualities like indivisibility and jointness of supply that make them inherently public goods (Olson 1965, 14–16). To institutionalists, *public* and *private* refer not to intrinsic qualities of phenomena but rather to a specific type of social institution through which people relate to, or use, the phenomenon. McCay (1996, 12) makes this point clearly: "One should distinguish between features of the resource and those of the ways people choose to relate to the resource and to each other."

Global Epistemic Institutions

According to Robert Kohler (1982), scientific disciplines are institutions, in that they provide a social infrastructure within which people produce knowledge. Roles that scientists occupy, and the rights and responsibilities assigned to them, can be seen as mediating between producers and consumers of knowledge. The institutional context within which scientists work shapes the way they practice science.

Science is not the only epistemic institution, but it is a highly influential one.[3] A person occupies the role of scientist if he is recognized by members of an academic discipline as a member of the collective. An individual's epistemic role situates that person within the field. Within disciplines, scientists may form smaller in-groups based on shared beliefs about phenomena being studied, the appropriateness of explanatory models, acceptance of a particular research method, and so forth. The field of science and technology studies chronicles many such debates in the history of science.

In the widest sense, the role of scientist consists of a cluster of behaviors such as researching, publishing, and teaching. Members of a discipline (or in-groups within disciplines) decide for themselves what science means, what should be accepted as science, and who counts as a scientist. They control academic appointments, research grants, what gets published in journals, and the awarding of academic degrees (Polanyi 1956, 215). Within a discipline or in-group, work is guided by the epistemic frameworks or methodological rules that the group recognizes as valid.

Hybrid Institutions

Scientists and the State Hybrid institutions refer to areas where different types of institutions, for instance, the state, the market, and science, are intertwined. Scientists have sought political patrons, and vice versa, at least since the early modern era. Some of the most celebrated innovations, such as Galileo's heliocentric theory, were developed in spaces opened up by European court society. The relationships between the Royal Society and the British Navy, and the French state and the Académie, were forged and sustained in a climate of strategic competition (Gower 1997). Political and economic institutions proved mutually stimulating and mutually sustaining.

The sinews binding political and epistemic institutions tend to change over time. The role of sovereign, and the rights and responsibilities attached to it, are created by a group's expectations. In the United States the military has developed a web of ties with university scientists through defense advisory committees. The National Academy of Sciences

(NAS), developed in the middle of the Civil War, is a private foundation with a federal charter, created to provide expert advice to the government. In 1916, as the United States prepared to enter the war in Europe, the NAS created the National Research Council (NRC). The NRC was "to encourage both pure and applied research for the ultimate end of national security and welfare." To this end, it was "to promote cooperation among all the research institutions of the country." After World War II, the scope of science advice widened considerably (Kevles 1971, 108–112). President Eisenhower created the President's Science Advisory Committee in 1958, bringing many physicists into the hybrid network.

These networks widened substantially in the decades following World War II. Military research assumed large-scale proportions, supplying 20–25 percent of the budget for research and development, and employing 20–25 percent of the scientists and engineers in the United States (Smit 1995, 201). Each branch of the military developed its own network, for instance, the Defense Science Board, the Army Science Board, the Air Force Scientific Advisory Board, and the Navy Research Advisory Committee. The military sponsored an interdisciplinary group of advisers known as JASON and brought them together in regular summer sessions. The expanding web of advisory committees encouraged scientists to develop and to maintain hybrid roles.

When acting in a hybrid role, scientists tend to frame their research agendas to meet both political and epistemic requirements. Most commonly, individual scientists create contracts with the state for particular projects.

When one role is superimposed on another, it may entail contradictions among different sets of rights and responsibilities. For example, the state may choose to classify information and restrict its dissemination to authorized members of the group. This conflicts with the obligation scientists have to disseminate their research results. A scientist in an academic setting will find that her career suffers if she cannot publish research results. In other situations, researchers may be able to frame results in more than one way, to satisfy competing and contradictory institutional requirements.

Scientists and the Market A second hybrid domain involves overlap between epistemic and economic institutions. In this domain, the rights and rules that attach to the role of scientist overlap with those of the property holder role. This epistemic/economic hybrid is of European origin and dates to the sixteenth century. There is some evidence that natural philosophers sought to translate practical innovations into commercial gains. For example, Venetian records show a patent assigned to Galileo Galilei, covering a pump and water distribution system, dated 1593 (Prager 1944, 719). In Europe and throughout the West, epistemic and economic institutions developed in tandem, each reinforcing the other.

The epistemic/economic hybrid has certain potential problems built in. There is an inherent tension between defense of private intellectual property rights and the dissemination of knowledge. Historically, inventors have claimed that patents prevented them from developing new ideas in technological areas dominated by large corporations (Etzkowitz and Webster 1995, 484). Academic scientists with commercial interests might be less likely to submit their findings for peer review, for fear that the innovations might be pilfered by peer reviewers who are also commercial competitors (505). More fundamentally, the question arises as to whether the patent allows the individual to draw concepts, knowledge, or ideas from the public domain and convert them to private property (Boyle 1996, 50; Lessig 1999).

Despite built-in tensions, this hybrid domain has expanded rapidly in the past two decades. Academic entrepreneurs are becoming common in a number of fields, in which scientists are interested not only in symbolic recognition of their contributions, for example, through citation (Latour and Woolgar 1979) but also in marketing their work. This has been especially true in the areas of microelectronics, computers, and biotechnology (Etzkowitz and Webster 1995, 480–481).

Scientists, the State, and the Market The political, economic, and epistemic institutions of concern date to the early modern era and emerged together as mutually reinforcing domains. In the sixteenth century, Galileo was simultaneously a scientist, a beneficiary of the patronage of the Medici, and a holder of at least one patent granted by the city-state of Venice (Prager 1944, 719; Gower 1997). In seventeenth-century England, members of the Royal Society developed a synergistic relationship with the Royal Navy. British sea power, in turn, extended the commercial fortunes on which Society members relied for support (Drayton 1998). In the late twentieth century, the development of information technologies has been powerfully shaped both by sovereign competition and by the marketplace.

2.2 Some Mechanisms for Fixing Social Beliefs

Global political, economic, and epistemic institutions can be seen as complex systems. The question is whether and how these systems shape the generation and use of new knowledge. This is important because societies often must form beliefs quickly (e.g., about a particular threat) before selecting among alternative actions. On the other hand, uncertainty about matters of fact can break down the process of intentional group action. Two related questions present themselves: How (through what mechanisms) do institutions shape the generation of knowledge? and How (through what mechanisms) do groups fix beliefs? Again, *fix* is used to indicate that the mechanism is used *to establish* belief and *to repair* uncertainty. Based on the case studies in part II, three mechanisms appear to be most salient: the positional fix, the statutory fix, and the committee fix.

Positional Fix

Social roles such as the role of committee chair, position an agent with respect to institutional structures. Roles specify what an agent should do, or ought to do, in certain social contexts. With respect to institutions, roles are typically formal and professionalized, and have specific rights, responsibilities, or rules attached. Although rights and rules do not determine a precise range of possible actions a role holder may take in various situations, they provide guidance as to what the role holder should do, given the purpose of the role position in the institution and in society. To take a simple example, the role of scientist typically involves researching, publishing, and teaching.

Roles are generative in the sense that in new or unfamiliar situations agents use them as a guide to action. When employing a positional fix, agents use their role position as a guide to framing research or accepting beliefs. In other words, the agent uses the role as a guide (or frame) when generating new statements or deciding which belief to accept. For example, this happens when a scientist refers to his role as chair of a committee (and the rights and responsibilities charged to the committee) when planning his research.

The example becomes more complex when an individual accepts more than one role at the same time. For example, the Director of the Scripps Institution was responsible (at least in part) for planning the organization's research agenda. Before World War I, the Director's role was primarily epistemic, relatively free from political and economic responsibilities. However, shortly before World War I, the Director became a member of the National Research Council, which entailed certain obligations related to national defense. Thereafter, the Director responded to a blend of epistemic and political responsibilities when planning research. The social role became a mechanism through which political changes (like the emergence of U.S. global hegemony after World War II) shaped the Scripps Institution's research program.[4] There are numerous other examples of such hybrid roles, combining political roles (a position on the NRC, a Department of Defense or Joint Chiefs of Staff advisory committee) or economic roles (entrepreneur, member of a corporate board) with epistemic ones.

Statutory Fix

A second mechanism is the statutory fix. Institutions often embrace certain modes of thought and thereby privilege certain ideas while passing over others (Ruggie 1983). Regimes tend to embed particular ideas, which in turn shape policy outcomes (Goldstein and Keohane 1993). Ideas (or frameworks) are embedded in a variety of formal and informal institutions. The statutory fix refers to use of such embedded ideas to shape the generation of new statements or acceptance of belief.

For example, the Director of the IATTC referred to the Convention for the Creation of an Inter-American Tropical Tuna Commission (a treaty signed in 1949) for broad guidance when designing the organization's research. There are a number of ways the Director could have organized it, for instance, as a general survey of all organisms in an area, as a model of predator-prey interactions, and so forth. The treaty that created the IATTC embeds maximum sustainable yield (MSY) in Article II, specifically authorizing research for the purpose of informing regulations to maintain MSY. Because this particular framework is embedded in the treaty rules, and because performing his job requires the Director to fulfill responsibilities laid out in the treaty, the embedded idea, MSY, emerges as the dominant epistemic framework guiding the research. The statutory fix is a second pathway through which the institution shapes the generation of new knowledge about the ecosystem. Further examples can be drawn from other treaties that embed particular epistemic frameworks: the World Trade Organization Agreements (liberalism); the UN Convention on the Law of the Sea, Article 61 (conservation); the Convention on Biological Diversity, Article 8 (d), (f), (h) (ecosystems approach).

Committee Fix

As explained in section 2.3, formation of beliefs can be seen as acceptance, either by an individual or an organization (Tuomela 2000b). Institutions enter in when people develop rules and social practices for generation and acceptance of particular kinds of beliefs on an ongoing basis. Many organizations have more or less standardized practices for forming beliefs as a group. The committee fix refers to precisely this type of mechanism.

For example, each year the IATTC formally accepts certain beliefs about the abundance of tuna and dolphin stocks, in order to guide regulation. It forms a belief about whether yellowfin stocks are fished below, at, or beyond the point of MSY in a particular season. This matters because changes in the organization's beliefs—assuming that the interests of stakeholders do not change—point to different regulatory actions.[5] If fishing poses no threat (is below MSY), then no regulation is needed. If fishing does pose a threat (is at or beyond MSY), then some regulation is required. Other examples can easily be found: a group of central bankers forming beliefs about inflation rates (e.g., in order to set interest rates); Working Group I of the Intergovernmental Panel on Climate Change (IPCC) forming a belief about the anthropogenic threat to the climate system (to ground policy recommendations); the U.S. Joint Chiefs of Staff forming a belief about the threat from another state (to ground military actions).[6] The committee fix refers to rules and regularized practices through which groups like these generate new knowledge or accept particular beliefs.

2.3 Social Knowledge

So far, analysis has proceeded using the working definition from chapter 1, knowledge as accepted belief. It is necessary to specify more precisely how the term *knowledge* is used.

Beliefs in Individuals and Groups

In a narrow sense, a belief is a mental event, an idea in the mind of an individual. This is the most common use of the term in international relations (Yee 1996; Goldstein and Keohane 1993).

But in a wider sense, *belief* can refer to a linguistic proposition an agent is willing to accept. Beliefs can be about matters of fact (["I believe] the water is poisoned"). Beliefs can also be normative, that is, they can express should/ought statements (["I believe] you should not drink poison"). Crucial to this working definition of knowledge is acceptance, in the sense that the speaker has committed (if only temporarily) to the truth of the statement.

Specifying knowledge this way opens the possibility that groups as well as individuals can accept beliefs. Groups can accept beliefs in a sense that is irreducible to the beliefs of constituent individuals (Tuomela 1995). This is not to say that group beliefs are *never* reducible to individual beliefs but rather that they are not *necessarily* reducible. A group may establish a mechanism whereby individual members express beliefs, and through some decision rule (three fourths in favor, 51 percent in favor) the group forms a belief. This mechanism can be called aggregative in the sense that one "adds up" individual beliefs to arrive at a group belief. Group beliefs can be, but are not necessarily, formed from the bottom up.

It is also possible that a group establishes one or more beliefs without polling its members (Tuomela 1995). The Catholic Church, for example, believes that miracles occur. Further, it is possible (if unlikely) that a group accepts a statement that none of its members accepts (in the narrow sense). For example, it is the official belief of the Communist party in country X that "capitalist countries will soon perish." As a party member, an individual may be obliged to accept the statement even if as an individual he does not believe it (in the narrow sense).

The point is not that group beliefs and individual beliefs differ all or even most of the time. Rather, in principle, one can identify social beliefs that are irreducible to the beliefs of constituent individuals. The question becomes, How (by what mechanisms) do groups form beliefs in the first place? The task is to identify institutionalized, or standardized, practices that groups or societies have developed to establish social beliefs.

Accepted Belief

Belief in the wide sense (accepted belief) differs from belief in the narrow sense (an idea in the mind of an individual) in the following ways. First, acceptances are voluntary and tend to involve some exercise of will. Beliefs as interior mental events do not. For example, imagine a boy and a girl who are very much in love. The boy falls out of love with the girl. The girl has difficulty believing it (in the narrow sense): she can't get it out of her mind that in his heart the boy still loves her. But, she decides to go on with her life, and so she accepts that he no longer loves her (Tuomela 2000b).

Similarly, consider a scientist who is nagged by doubt about a particular scientific finding. At some point, if she is to get on with her research, the scientist must decide whether to accept the belief or to reject it. The scientist may not believe it (in the narrow sense), but to get on with research, she accepts it.

Second, acceptances are an all-or-nothing matter, whereas beliefs (narrowly defined) come in degrees. Again, consider the scientist. She may be ridden by doubt. But she *must* decide whether to accept a given finding, or her work will be paralyzed. At some point in time, the scientist either accepts or rejects the finding, and at this point the decision is all-or-nothing. Yet the acceptance is also temporary and subject to future revision in light of new evidence.

Third, acceptances concern propositions expressed linguistically or symbolically in the form of charts, graphs, tables, and so on. Beliefs (narrowly defined) are not necessarily expressed this way. Beliefs formulated as linguistic and symbolic propositions are observable. For example, the IPCC believes that "projections of future global mean temperature change and sea level rise confirm the potential for human activities to alter the Earth's climate to an extent unprecedented in human history" (Houghton et al. 1995, xi). To study the IPCC's beliefs, one need not look inside the heads of each of its members. Rather, one can study the group's beliefs once they have been accepted as one or more linguistic or symbolic propositions. In short, *belief* refers to propositions that one or more agents accept. *Knowledge* refers to one or more accepted beliefs. The question is whether and how institutional mechanisms shape the generation and use of knowledge.

Specifying knowledge as accepted belief is useful because statements are observable whereas ideas in individuals' minds are not. Limiting the focus to scientific knowledge, the analyst can observe knowledge produced in the form of published papers; reports, bulletins, or working papers of scientific organizations; and published manuscripts. Statements may appear in linguistic form, in written text. They may also appear in symbolic form, in tables, graphs, charts, or maps.

2.4 Methodology

So far, this chapter has outlined a theory of political, economic, and epistemic systems as overlapping sets of institutional rules. It suggests that a complex set of social roles, rights, rules, and norms shapes human action, including the generation of new knowledge. The claim is that certain mechanisms—the positional fix, the statutory fix, the committee fix—shape the generation of new knowledge. In order to apply this theory in a meaningful way, it is necessary to spell out precisely how its implications differ from possible alternatives, such as neorealism and the interest group approach.[7]

Case Selection

The cases in part II have been carefully selected to enable the reader to differentiate the institutional argument from the neorealist and interest group approaches. In chapters 3–5, the history of the Scripps Institution from about 1905 to 1970 is used to evaluate the institutional argument as compared to the neorealist one. Chapter 6 uses the history of the IATTC from 1950 to about 1998. This allows the analyst to consider international structure (as the neorealist defines it) constant, and to consider the effects of variation in interest groups.

Once again, the problem of concern is that groups of people often must act despite a context of uncertainty. Is a tornado about to strike the town? What is its path? In emergency situations, societies must establish some belief about the nature of the threat in order to act quickly. More routinely, it is necessary to establish certain facts—the state of the economy, the quality of an ecosystem—as a basis for setting regulations.

The institutional argument is that groups use certain mechanisms to form beliefs. They are the positional fix, the statutory fix, and the committee fix. Through these mechanisms, institutions shape the generation of new knowledge, often over long periods of time. Once formed, accepted beliefs repair uncertainty (at least temporarily), illuminate interests, and point to preferred actions.

Alternative Explanations

Neither neorealism nor the interest group approach has developed a theory of knowledge generation. In the past, neither approach has considered ideas or beliefs to be particularly important determinants of human action.[8] Typically, beliefs are specified exogenously, as if the world were transparently knowable to the agents being modeled. Since neither approach has devoted much research to the topic, any discussion of alternatives risks setting up straw men.

Nevertheless, it is necessary to differentiate the present approach from possible alternatives by exploring what the standard approaches would predict, were they to take up the problem. In general, critics have responded as follows. People consider alternative options for action, for instance, producing knowledge of type A vs. producing knowledge of type B. They then consider interests, defined as the payoffs to alternative options. Interests determine the kinds of knowledge people produce and the beliefs they accept. In situations in which the state or private industry funds research, the interests of the principals dictates the type of research funded.[9]

Institutionalism vs. Neorealism

The neorealist claim would be cashed out as a structural theory. First, according to neorealism, powerful states act to maximize power (or security). The most important factor influencing the actions of states in the international system is international structure, an ordering of states based on capabilities (Waltz 1979; Mearsheimer 1994; Krasner 1999). As international structure changes, say, from bipolar to multipolar, we would expect to see large-scale changes in patterns of state behavior. Similarly, as a state moves from one position in the system to another, say, nonhegemonic to hegemonic, we should expect to see changes in its behavior.

In brief, the institutional approach predicts that the Scripps Institution would generate defense-related research when the United States accepted a role as great power, but not otherwise. By contrast, the neorealist approach predicts that the Scripps Institution would generate defense-related research when the United States ranked as a hegemon (based on material capabilities).[10] Prior to World War II the predictions differ sharply: the United States had the capabilities of a regional hegemon from about 1890 until about 1945. Therefore the neorealist approach would predict that the United States would fund research related to its interests, such as military applications of marine science, from about 1890 on.

From an institutional perspective, the United States accepted the role of great power during World War I, but not before (1900–1916) or after (1920s and 1930s) (Carr 1939; Wight in Toynbee 1952; Hillman in Toynbee 1952; Bull 1977, 212; Kennedy 1988, 328; Gaddis 1997, 6–7). Therefore institutionalism predicts that the United States would fund military applications of marine science when it accepted a role as great power (or hegemon), but not when it rejected the role. As applied to the period prior to World War II, the neorealist and institutional approaches make different predictions. After World War II, when the United States emerged as a global hegemon, the predictions are the same.

Institutionalism vs. Interest Group Approach

A second alternative is the interest group approach. According to this view, institutions don't matter. Rather, the relative strength of interest groups in a particular research area determines what gets done. If fishing firms require certain kinds of data to do business, they hire scientists and specify what frameworks they must use. Alternatively, if firms are organized and lobby governments, scientists working in the public sector produce the kinds of data firms need to conduct business. "He who pays the piper"—the interest group—"calls the tune."[11]

In chapter 6, the focus on the history of the IATTC allows the analyst to home in on the importance of interest groups, including firms and environmental activists. The institutional argument is that through the statutory fix and the committee fix, the IATTC formed particular beliefs about the status of fish and dolphin stocks. As these beliefs change, the IATTC's regulatory actions tend to change, other things equal.

The predictions of the interest group approach differ sharply. The interest group argument is that the relative strength of interest groups (over time) determines the particular beliefs the IATTC accepts. For example, if it is in the industry's interest for the IATTC to form the belief that stocks are abundant, the IATTC will do so. Furthermore, the interest group approach predicts that the relative strength of interest groups over time determines the organization's regulatory actions. For example, if firms want the IATTC not to regulate, it will not. Over time, if environmental activists become more powerful, pressing the IATTC to tighten regulations, the organization will do so.

The IATTC's history is important to the present study primarily because it allows the analyst to hold the neorealist argument constant while focusing in on the interest group alternative. In other words, neorealists argue that the state's interests (dictated by the international power structure) determine funding for research. In neorealist theory, the international structure was bipolar from 1950 until about 1989. Therefore, international structure (which is constant) cannot explain changes in the IATTC's beliefs about the status of yellowfin and dolphin stocks, which changed considerably between 1950 and 1990. Nor can it explain changes in the IATTC's regulatory actions in the same period.

The history of research at the Scripps Institution and the IATTC can be seen as hard cases for institutionalism.[12] Scripps' history (1905–1970) includes significant variation in international political structure (whether defined in institutional or neorealist terms). Interest groups were not particularly active. Because oceanography in the post–World War II era was so penetrated with funding from the state (Mukerji 1989), the case seems likely to confirm a neorealist interpretation. That is, changes in international political structure (as neorealists define it) determined the direction of changes in knowledge.

Considering the history of the IATTC (1950–about 1998) holds power structure constant to tighten the focus on interest groups. Because fishing firms were involved in the IATTC's history from the beginning (Scheiber 1984; Parker 1999), the case seems most likely to confirm an interest group interpretation. That is, changes in the relative strength of interest groups over time determined the direction of changes in the IATTC's knowledge and thereby also the changes in its regulatory actions.

What Is Observed

The positional fix refers to an agent's use of a social role, with attached rights and rules, as a guide when framing research or selecting beliefs. In the case studies, observations are made as follows. A primary social role is identified as a cluster of rights and rules (or responsibilities) that positions an individual with respect to a group. The individual's description of these rights and rules can be found in letters, memoirs, and annual summaries of work completed. Similarly, descriptions by others in the group (or those in a position to observe the group) can be found in primary source documents. Change is observed when the rights and rules are significantly altered, for instance, when political responsibilities are added to epistemic ones. Change can also be observed when an individual accepts a role he had previously rejected (or vice versa).

Secondary social roles are identified as a cluster of rights and rules that positions a group with respect to other groups. Primary source documents contain explicit references to them (in interviews with participants, letters, professional papers). A change in a secondary social role can be observed when there are explicit changes in the rights and rules attached to a role. Change can also be observed when a group accepts a role it had previously rejected (or vice versa).

The statutory fix refers to use of ideas, embedded in formal or informal rules, to establish belief or to frame research. Ideas embedded in rules can be identified with a review of constitutive legal documents. Of interest to the present work are epistemic frameworks, approaches to a subject matter within which a number of models or theories can coexist. For example, Article 61 of the UN Convention on the Law of the Sea refers to conservation (an approach that includes different types of management models, including MSY).[13] Article 8 of the Convention on Biological Diversity refers to the ecosystems approach, an approach that encompasses a variety of models (Golley 1993).

The committee fix can be observed by identifying regularized practices through which groups accept beliefs. These may include annual or quarterly meetings, the goal of which is for the group to adopt a research report or a synthesis report. To count as institutionalized, the practices for accepting beliefs must be regularized and carried out in a similar fashion over a discrete period of time.

To observe knowledge (the knowledge of an individual or a group), the analysis focuses on accepted beliefs. To count as knowledge, it is necessary that the statement be accepted (by the individual or group) and that there be evidence of individual or joint acceptance. Evidence can include peer-reviewed statements published in books and journals, annual reports, and working papers. (Peer review is a form of acceptance by a community; publication in annual reports constitutes group acceptance.) Statements may be in verbal form or in digital form (published in tables, graphs, charts, or maps). Changes in knowledge can be observed when there are changes in beliefs about a phenomenon ("yellowfin are scarce this year; last year they were abundant"). It can be observed when there are changes in the kinds of questions asked, the topics explored, and the disciplines, approaches, or models used to answer these questions. For example, a shift from ecosystems modeling to undersea acoustics counts as a shift in the kinds of knowledge produced.

Whereas chapters 3–5 focus on more broadly based changes in knowledge, chapter 6's structured case study allows for a more nuanced analysis. In this case, changes in knowledge refer to changes in beliefs about the status of yellowfin and dolphin stocks. Depending on the belief accepted, the degree of threat is coded from low to high. Changes in regulatory action refer to changes in annual quotas, or limits on incidental killing of dolphins, that the organization used to limit the adverse impact of the fishery. Regulatory actions are coded from weak to strong. The emphasis is on mechanisms, including the statutory fix and the committee fix, that the organization used to generate certain kinds of knowledge. As the organization came to believe that the anthropogenic threat to marine life had increased, its regulatory actions were strengthened as a result.

2.5 Summary

This chapter has outlined a theory of political, economic, and epistemic systems as overlapping sets of institutional rules. Their depiction as complex institutional systems departs from, and extends, standard regime theory in international relations. These institutions give rise to a complex set of social roles, rights, rules, and norms that shape human action, including the generation of new knowledge. The chapter also introduced three institutional mechanisms—the positional fix, the statutory fix, the committee fix—that mediate the generation of knowledge. These mechanisms are only one part of the conceptual landscape outlined in chapter 1, but they are the mechanisms that are most salient, given the case studies (chapters 3–6). Finally, borrowing from recent work in political philosophy (Tuomela 1995; 2000b), the chapter specified knowledge as accepted belief and clarified the sense in which groups (organizations or committees) can be said to accept beliefs. The following chapters use historical evidence to show how these institutional mechanisms shaped the generation of knowledge at the Scripps Institution and the IATTC.

II

Scripps Institution of Oceanography and the Inter-American Tropical Tuna Commission, 1900s–1990s

3

Exploring Pacific Ecosystems: Scripps Institution, 1905–1917

Emergency situations—war, collapsing fisheries—are frequently clouded by uncertainty about matters of fact. People must generate some knowledge about the threat before they can decide how to act. Typically, broad patterns of research activity have been established long before the emergency arises. As a consequence, it is important to investigate the mechanisms through which these broad patterns of research activity are generated as well as the mechanisms by which groups accept some beliefs in order to ground action. In this and the next two chapters, I illustrate how the positional fix provided the mechanism by which the knowledge generated at the Scripps Institution was influenced by various institutional factors. The role of Director provided the mechanism for establishing the epistemic framework that guided research at Scripps as well as the mechanism through which changes in global institutional configurations were translated into changes in the types of knowledge generated at the Scripps Institution.

This chapter lays a baseline from which future changes are measured. It documents how W. E. Ritter established a role for himself as Director of the San Diego Marine Biological Station (which was renamed the Scripps Institution for Biological Research in 1912 and the Scripps Institution of Oceanography in 1924).[1] The need for a research director arose from Ritter's proto-ecosystemic view of what the marine biological station's work should be. Construction of the role of Director (explanatory variable) shaped the knowledge generated at the Scripps Institution (dependent variable). As the role changed over time (documented in chapters 4 and 5), so too did the kinds of knowledge Scripps generated.

The chapter is organized as follows. Section 3.1 presents the political and economic contexts. Section 3.2 introduces the Scripps Institution's program, set in the context of marine science of the early twentieth century. Ritter's role position can be seen as a single observation (on the explanatory variable), with change over time presented in later chapters. Section 3.3 describes the research program in these early years, from about 1905 until about 1915. This section can be seen as a single observation (on the dependent variable), with changes over time documented in later chapters. Section 3.4 argues that the institutional approach better accounts for research activity than the alternatives.

3.1 Political and Economic Contexts

Global Political Role: None

The Scripps Institution's earliest work was not tied to the national security establishment, although the United States had by the early twentieth century established itself as a hegemon in the Western Hemisphere and in the Pacific region. By any measure, by 1890 the United States had the brute capabilities to act as a great power, but it did not accept an active (secondary) role in European politics.[2] In the 1890s, Congress funded construction of battleship fleets intended to support U.S. dominance over the western Atlantic, the Caribbean, and the eastern Pacific, and the European powers recognized this new naval power (Sprout and Sprout 1946, 211; Kennedy 1988, 195). By 1907 the United States had the third largest Navy in the world, measured in tonnage, behind Great Britain and Germany (Kennedy 1988, 203, 247). In terms of combat capability, because a large proportion of the U.S. ships were of the newer Dreadnought type, the U.S. Navy was second only to Britain's (Sprout and Sprout 1946, 272).

At the same time, the U.S. economy became intertwined ever more tightly with Europe's. Because the international economy operated on a gold standard, U.S. trade surpluses consistently brought in a flood of capital from Europe. The U.S. Treasury accumulated nearly one third of the world's stock of gold at this time (Kennedy 1988, 245). By the turn of the century, the U.S. economy was "already becoming a vast but unpredictable bellows, fanning but also on occasions dramatically cooling the world's trading system" (245).

After a brief bid for colonies in the Pacific under Theodore Roosevelt, the United States rejected what could have been an active role in European politics. President Taft (in office 1909–1913) largely returned to the more traditional U.S. policy of isolation from European international politics (Sprout and Sprout 1942, 286). The United States maintained "absolute neutrality" and "complete disinterestedness" in a number of European conflicts, including the Moroccan crisis, the war between Italy and Turkey, and the Balkan War (287). Although the United States had the brute capabilities of a great power, Americans rejected the role.[3]

> With a navy second only to that of Great Britain, the goal toward which Roosevelt had worked, the U.S. could play a strong if not decisive role in protecting either of these powers [Britain and Germany] from the other. While such considerations received some public attention [before World War I] . . . they were not cited . . . in any public utterance of the Administration, which chiefly supported its policy of keeping pace with European naval development on the ground that European navies constituted a potential standing menace to the territorial security of the U.S. and to the Panama Canal.[4]

Military and economic strength created the potential for the United States to assume a major role in European international politics, but it did not choose to accept it.

President Wilson continued this policy of isolation through the winter of 1914–15. In his annual message to Congress, Wilson declared that the United States remained "at peace with the world" (Sprout and Sprout 1942, 318). Public opinion polls supported continued disengagement (321). It was not until a German U-boat torpedoed the British ship *Lusitania,* killing one hundred Americans, that Wilson sent official protests to the German government. When the United States entered World War I, it assumed a role in the alliance against Germany.

Economic Role: None

Just as the Scripps Institution began without obligation to the military, it was also free from commercial demands. The United States had not yet asserted claims to ownership of resources on the continental shelf or elsewhere in coastal waters beyond three nautical miles (Hollick 1981). In southern California at this time, marine life adjacent to the Pacific was used on an open-access basis. There was little if any demand for marine scientific research in support of increased fishery yields. Overexploitation—with the exceptions of fur seals and Bristol Bay salmon—was not yet a problem (Burke 1994, 4–5; Hollick 1981, 20, 27). Major fisheries in southern California, tuna and sardines, were generally less intensively exploited than those in the North Atlantic.

The absence of commercial development in California at the time (1890–1915) was very different from the situation in Europe. For example, the Kiel school, beginning in the 1880s, took up the problem of failure of north German fisheries (Mills 1989, 4–5).

The periodic failure of herring fisheries in northern Europe was a major stimulus to European governments' creation and support for the International Council for the Exploration of the Sea (ICES). Pressure from fishing communities also stimulated government support for marine science at the Plymouth and Port Erin stations, and at Spencer Baird's laboratory in Woods Hole.

At the Scripps Institution, by contrast, the Director's role did not include any such economic responsibilities. As Charles Kofoid, Assistant Director of the San Diego station, told Ritter in 1908, oceanographers at European laboratories were enthusiastic that the San Diego station could design its research program with such "freedom from economic pressure."[5]

3.2 The Need for a Director at the Scripps Institution

To explain why research in the early years required a Director, it is necessary to explain Ritter's ecological framework in some detail. As Ritter saw it, to study marine ecology properly required some degree of central coordination. A Director was needed to plan, organize, and oversee the research. This is not to say that individual lines of research were driven out: they were not. But as far as the main program at the Scripps Institution is concerned, the research agenda required a holistic design, and this requirement gave rise to a (primary) social role of Director.[6]

Ritter's Framework: Holistic and Ecological

Ritter described his early vision for the Scripps Institution's program as follows:

> When viewing this whole field of knowledge, and the means and methods of investigation, one must be struck by the prevailing uniformity and inadequacy of the existing marine stations for coping with the situation. . . . They have been and are, with few exceptions, primarily resorts for individual investigators of specific biological problems, and not for systematically attacking the problems of marine biology proper.[7]

His wife, Mary Ritter (1933), also understood the work to be different from most marine biological laboratories of the day. She observed, "The common idea of a biological research laboratory was a place where trained persons could go to obtain facilities for individual work, entirely unrelated to any other work in the same laboratory. Mr. Ritter's idea was to lay out fields of work covering as many problems . . . as possible and to employ specially trained workers to solve these problems, according to the usual custom in astronomical observatories" (266–267).

In Ritter's day there were a number of precedents for large-scale, organized programs in marine science. Even before the *Challenger* expedition of 1872–1876, scientists had explored marine life in the Pacific in a number of large-scale expeditions. For example, Thomas Henry Huxley surveyed the Pacific in British Naval vessel *Rattlesnake* between 1840 and 1850, collecting deep sea organisms (Schlee 1973, 95–96; Deacon 1980, 229). In 1868, Charles Wyville Thompson and William B. Carpenter had surveyed the Pacific in the British naval vessel *Lightning*, taking observations of temperature distributions and sampling life forms in the deep sea (Deacon 1980, 306–309; Mills 1980, 360). From these and other expeditions, Ritter had reason to believe the Pacific Ocean held a wealth of life forms yet to be discovered.

Intellectual Influences on Ritter

Joe LeConte and John Muir Unlike most British and American zoologists of his day, Ritter thought about the Pacific and life in it as a single complex organism. To explain how Ritter came to think in holistic terms, it is necessary to consider how he was influenced by Joseph LeConte, a professor of geology at the University of California at Berkeley, and by John Muir, the famous naturalist. These men impressed upon Ritter the importance of studying ecological wholes and the need to understand the complexities of interdependent parts.

Ritter had moved to California from Wisconsin in 1885 to study as an undergraduate with LeConte. "Uncle Joe" had earned a devoted following among his students, and organized hiking expeditions in the High Sierras.[8] He was one of a small group of men who helped Muir organize the Sierra Club (Gifford 1996, 307; Cohen 1988, 19).

Unlike LeConte, Ritter was not involved in Muir's struggles to protect wild places. But Ritter did take part in the hiking culture that grew up around them. He found time to join an expedition in the spring of 1898. In a letter to his uncle from the Yosemite Valley, Ritter describes it as a "wholly different order of glory."[9] LeConte drew forth Ritter's sense of wonder about the natural world, a sense that pervaded his work throughout his career.

Ritter himself met Muir on the Harriman Alaska Expedition in the summer of 1899. The expedition was organized by railroad magnate E. Henry Harriman as a survey of Puget Sound. Harriman brought artists and naturalists together with experts in geology, botany, forestry, and zoology. Muir by that time was already an icon to those who wanted to protect California's wild places (see Cohen 1988; Nash 1982).

Their meeting was particularly memorable for Ritter, who recorded it in his journal: "Spent balance of day in ramble over the hills . . . [hoping] to learn something of the birds and their songs . . . came upon Muir [and others] . . . high up the hill—1400 feet. One of the most beautiful afternoons I have ever spent."[10]

Muir's memories of the trip reflect a similar spirit: "No doubt every one of the favoured happy band feels, as I do, that this was the grandest trip of his life. . . . I hope to have visits ere long from Professor Gilbert . . . and Earlybird Ritter."[11]

In October, Muir wrote to Ritter: "My dear Prof. Ritter, Come when you can—anytime and let's talk over our grand novel raid into the fine fruitful northern Wilderness. Keeler, Cap. Doran, Merriam, Gannett, and Gilbert have been here, and now of course you must come."[12]

Ritter's friendships with LeConte and Muir reflect his general orientation to biology in a spiritual and a philosophic sense as a "science of life."[13] It was Muir who wrote, "When we try to pick out anything by itself, we find it hitched to everything else in the universe."[14] But with slight amendment, it could have been Ritter: *when we try to pick out anything by itself, we find it hitched to everything else in the ocean.*

Alexander Agassiz and Mainstream Marine Zoology Another influence on Ritter's thought pulled in a different direction. That was Alexander Agassiz, who a century ago was the most prominent marine scientist in the United States (Murray 1911). As a graduate student, in the summer of 1890, Ritter worked in Agassiz's lab in Newport, Rhode

Island.[15] Agassiz's expeditions in the Atlantic on the *USS Blake,* and in the Pacific on the *Albatross,* generated massive collections of marine organisms, many of which were donated to Harvard's Museum of Comparative Zoology (Murray 1911, 880; Zinn, in Sears and Merriman 1980). Researchers collected organisms, then brought them to the lab for dissection and description.

Agassiz encouraged Ritter's work. He visited La Jolla in 1905 and donated books and scientific equipment to help get the program started.[16] Ritter remained in regular contact with Agassiz until the elder man's death in 1910. But insofar as he developed a holistic, ecological program, and studied the interrelationships between organisms and their environment, Ritter broke with Agassiz's more traditional methods of practicing zoology.[17]

European Models: Coordinated Yet Specialized Ritter looked not to the East Coast, but rather to Europe, to find models for his program. Among the more important influences were the Kiel Commission in Germany, the Port Erin station in England, and the hydrographic program organized under the auspices of ICES.

The Kiel school had pioneered quantitative techniques for studying plankton. Victor Henson, working at the University of Kiel on the Baltic Sea, initiated plankton studies beginning in the 1860s. Henson developed a technique in which researchers screened ocean water in a very fine meshed net, and examined samples with a microscope to count the number of tiny organisms in a sample. From this, they estimated the number of organisms per square meter of sea water.

Henson's colleagues at Kiel were inspired by his methods. Soon after, the German government established a Commission for the Scientific Study of the German Seas at Kiel. This stimulated a core group of zoologists, chemists, and botanists to further develop statistical sampling techniques. Beginning in 1902, the Kiel Commission became involved in collaborative work organized by ICES (Schlee 1973, 231–233). Ritter's program on plankton, in terms of design and methodology, was inspired by Henson's work at Kiel.

Ritter's plan for the San Diego station was also influenced by work at the laboratory of the Liverpool Marine Biological Committee, with its laboratory on the Isle of Man (established in 1892).[18] This laboratory was founded by W. A. Herdman, a professor of zoology at Liverpool. Herdman's group was itself patterned after work at Kiel. Herdman's program at Port Erin was structured holistically, in the sense that biological, chemical, and physical observations were coordinated in time and space. They were at the same time specialized and quantitative, to facilitate statistical analysis of the interrelationships between the parts. The Port Erin station investigated not only fish biology but also the distribution and abundance of plankton, hydrography, and invertebrate morphology (Mills 1989, 200).

Ritter also looked to ICES' hydrographic research program as a model for integrating discrete lines of research into a single biologically oriented program. These hydrographic studies began in the North Atlantic in 1902. ICES' overall goal was to identify why stocks of fish, particularly cod and herring, fluctuated so dramatically from season to season. Many scientists suspected that the fluctuations had something to do with fish migration patterns, which in turn had something to do with the temperature and circulation of water masses. The program had two parts, hydrographic and biological.

The hydrographic work in particular, aimed at understanding ocean circulation, required a single design. Observations—temperature, salinity, currents—had to be taken at the same time over a particular area. If the overall shape of the hydrographic program's research design changed, so too would the work of the individual chemists and physicists. Ritter used these programs—Kiel, Port Erin, and ICES' laboratory at Christiania (Oslo)—as models when he designed the program at the San Diego marine biological station.

Programs That Were Not Models Not all marine biological stations had a single design. At a number of other stations, the role of Director was less central to the conduct of the research. In other programs, like the Stazione Zoologica in Naples, and the U.S. Fish Commission in Woods Hole, Mass, the Director's role would have had less impact on the overall program.

Although he admired the Stazione Zoologica, founded by Anton Dohrn in Naples in 1873, Ritter wanted to avoid some aspects of its organization. The Stazione was a laboratory that provided facilities to visiting scientists.

Dohrn provided tables or work areas in his laboratories to visiting scientists, for a fee. Income was used to equip the lab and to operate collecting boats. Researchers from a variety of governments, universities, and scientific societies used the facilities (Schlee 1973, 69). Ritter himself had worked there in 1894, during a year of study in Europe.[19]

Spencer Baird, Director of the U.S. Fish Commission, designed his laboratory at Woods Hole, on Cape Cod, Massachusetts, after the Stazione. Researchers visited the laboratory during the summer, collecting specimens and dissecting them in the lab. Among them were zoologists from a number of East Coast universities, including Harvard, Johns Hopkins, Princeton, and Williams (Schlee 1973, 71). The U.S. Fish Commission operated a steamer, the *Albatross,* which Baird also made available to university researchers.

The biological work at these laboratories lacked the central coordination that Ritter wanted to establish for his. At the Stazione, and at Woods Hole, biologists were under no obligation to coordinate their work with one another. To the extent their findings formed a body of knowledge, it was an aggregation of particular findings.

The directors at these organizations can be seen as having a largely bureaucratic function and thus their roles as directors did not carry any specific epistemic responsibilities that would require them to organize or coordinate the type of knowledge generated at their laboratories.

Creating the Role of Director (Explanatory Variable)

The need for a Director with a specific epistemic role and attached responsibilities arose from Ritter's view of how marine ecological research should be conducted. One important element of Ritter's approach was holism. In a paper published in *Science* in 1911, Ritter argued that reductionism failed in the sense that specialized investigations of parts could not explain the organization of the whole. Functioning parts necessarily appear together with organization as a whole; therefore the part cannot be isolated as the "cause" of the whole.[20] His own framework was holistic, in the sense that the phenomenon of interest to Ritter was the ecology of the Pacific Ocean. Ritter thought of the ecology of the Pacific as a single complex organism.

Second, he believed specialized, quantitative lines of inquiry were necessary. He envisioned a survey of marine life in the Pacific, broken out into biological, physical and chemical lines. These specialized observations had to be coordinated in space and time in order to provide a picture of the whole.

Ritter's third core belief was that organisms should be studied in the context of their environment. The program was designed to survey the horizontal and vertical distribution of organisms, while at the same time collecting data on the physical and chemical properties of the ocean environment. His survey of the Pacific was intended to be primarily biological, but Ritter believed that to understand the ecology of the ocean it would be necessary to understand physical and chemical aspects of the ocean in which the organisms lived.

In each of these aspects, Ritter's framework anticipated ecosystems theory. The term *ecosystem,* coined decades later by Arthur Tansley, was holistic in the sense that a community or organism was the phenomenon of interest.[21] Although primarily concerned with ecological wholes, Tansley was anxious to maintain a connection with mechanistic, reductionist science in order to maintain ecology's reputation as a hard science (Golley 1993, 15). Tansley's emphasis on the interaction between the organism and its environment is regarded as an important conceptual advance in ecology (24).

Ritter's framework is perhaps best characterized as proto-ecosystemic. Unlike Tansley, Ritter did not fully develop a systems theory.[22] His concept of the ocean as an organism did not trace flows of matter or energy among parts of the system. Also unlike Tansley, Ritter did not develop a concept of systemic equilibrium. With LeConte and Muir, he shared an ecologist's holistic vision. Ritter's framework anticipated not only ecosystems ecology but also future work by Alfred North Whitehead and Aldo Leopold.[23]

In sum, in Ritter's view, to develop a holistic program of research in marine ecology, required central coordination. Therefore, he established a role for himself as Director. Attached to the role was the responsibility to provide a coordinated plan for ecological research, to assemble a team of researchers qualified in physics, chemistry, and biology to help implement it, and to oversee the progress of the research. In this way, the early program of research at the Scripps Institution was established as a group effort, with Ritter as Director and his proto-ecosystemic vision as the guiding framework.

3.3 Knowledge Generated (Dependent Variable)

Coordinated Research: Plankton Ecology

By means of his position as Director, Ritter coordinated and framed the knowledge generated at the Scripps Institution. The backbone of the station's program consisted of coordinated measurements by staff biologists, chemists, physicists, and hydrographers. In the early years, to take observations at sea, Ritter contracted with a local fisherman, Mr. Cabral, in his boat *St. Joseph*. Cabral ran his boat three days a week during the summer, and at regular intervals during the year. In 1905, Ritter participated in a research cruise organized by Alexander Agassiz, aboard the U.S. Fish Commission steamer *Albatross*. The station launched its first research vessel, the *Alexander Agassiz,* in the spring of 1908. By the end of its fifth year, Ritter and his team had identified 508 kinds of organisms, more than 100 of which were new to science (Raitt and Moulton 1967, 50).

By this time, Ritter had assembled a staff to carry on the station's program. It consisted of Ritter, as Director; John Dahl, captain of the *Alexander Agassiz;* a librarian; and two scientific assistants, Ellis Michael and Myrtle Johnson. Several scientists agreed to work part-time for Ritter's station. They included Professors Kofoid and Torrey of the University of California, C. M. Child of the University of Chicago, C. O. Esterly of Occidental College, G. F. McEwen and H. C. Burbridge, both of Stanford University.

McEwen, a physicist, joined the part-time staff in 1908. Part of his job was to develop a more comprehensive hydrographic survey of the waters off Southern California. McEwen coordinated his work with the biologist Ellis Michael, taking continuous hydrographic observations and plankton samples. The results were published as "Hydrographic, Plankton, and Dredging Records of the Scripps Institution for Biological Research of the University of California, 1901–1912" (McEwen and Michael 1913–1916). The body of work done by McEwen and Michael, relating variations in plankton abundance to hydrographic conditions, proved to be the core of the Institution's ecological work before the war. The station produced a snapshot of the marine ecology off the coast of Southern California and a basis for the future study of the marine environment.

Another part of the coordinated program involved observations of a form of crustacean, the *copepoda*. Esterly took observations of copepods in relation to other parts of the marine food chain. Statistical studies of copepods were coordinated with McEwen and Michael's observations of plankton and hydrography.[24]

Individual Projects

Ritter's role as Director did not influence *all* the knowledge Scripps researchers generated. In other words, Ritter's ecological program did not drive out all independent lines of research. A number of researchers worked part-time on the station's program and had time left over for their own individual pursuits. For example, McEwen began work on an upwelling near the coast of California. He used his own observations to test V. W. Ekman's theory of the effect of wind on surface waters. McEwen initiated statistical studies of sea surface temperatures and the climate close to shore. This work was not related to European work in physical oceanography, then being developed by the Bergen school. McEwen's large and complex mathematical formulations relating ocean temperature and currents were never published (Mills 1991, 259). He did, however, distribute climate bulletins, for a fee, to local power companies and farmers.

The research program advanced along a number of other lines, primarily biological. Ritter published a paper in 1906 describing the marine organism *Octacnemus,* based on specimens taken during the 1904-05 cruise on the *Albatross.*

Kofoid concentrated on microscopic organisms known as *Dinoflagellates,* identifying 350 species and 45 genera.[25] Kofoid (with assistance of Dr. Olive Swezy) published *The Free-Living Unarmed Dinoflagellata* (Raitt and Moulton 1967, 83). The staff biologist W. E. Allen published several technical papers on various phases of microphytoplankton.[26] Meanwhile, Francis Sumner, another staff biologist, developed a program to run tests on inherited characteristics in mice.[27] Ritter continued his work in philosophical biology, publishing *The Unity of the Organism, or the Organismal Conception of Life.*[28]

In 1912 the San Diego Marine Biological Association agreed to transfer Ritter's station to the university of California. Thereafter, the station was renamed the Scripps Institution of Biological Research. The university began to fund approximately half of the station's operations, the rest being from Ellen Scripps, and a small fraction from rental income from buildings on campus. The Institution continued to operate according to Ritter's original plan. Oceanographic research remained "primarily in the interest of, and subordinate to, the biological investigations."[29]

3.4 Alternative Explanations

Institutionalism vs. Neorealism

This chapter has argued that Ritter's role as Director shaped knowledge generated at the Scripps Institution (from about 1905 to 1915). By contrast, neorealism would predict that the United States would fund marine scientific research at times when the state had an interest in it. According to this approach, as the United States became a hegemon in the Pacific region (measured by its preponderant capabilities), the state would fund military applications of marine science. At this time (1905–1915), it is likely that such military applications would have included basic and applied research on submarine warfare.[30]

The facts do not support this interpretation. Prior to 1915, marine scientific research at the Scripps Institution was not funded by, or otherwise influenced by, the military. The evidence demonstrates that in the period in question (1905–1915), preponderant U.S. capabilities did *not* translate into militarization of marine scientific research at the Scripps Institution.

Institutionalism vs. Interest Group Approach

The interest group approach would predict that firms in Southern California would demand the kinds of research that would enable them to maximize profits. They would then pay scientists to conduct it. There is some limited support for this interpretation. For example, McEwen provided weather forecasts to local companies, for a fee. However, this research was *not* part of the Scripps Institution's main program. The evidence shows that interest group activity did not shape the Scripps Institution's coordinated program in plankton ecology at this time (1905–1915).

Furthermore, from the very beginning, Ritter's benefactors made no attempt to influence the content of research. In the early years, San Diego businessmen paid for Ritter's summer research.[31] These donors were not in the fishing business: they had no practical interest in marine science per se. Support for the research came without any specific obligations (Mary Ritter 1933, 259). This financial support enabled Ritter and his students to return to the San Pedro area in 1901, to rent and equip a bath house as a summer lab. In 1903, with Baker's help, Ritter raised more money to finance a temporary lab in the boathouse of the Coronado Hotel in San Diego. Ritter's financial backers formally created the San Diego Marine Biological Association and appointed Ritter Scientific Director. The program of research in its earliest years was funded by local businessmen, but the donors set no demands on the researchers.

E. W. Scripps and his older sister, Ellen Scripps, provided significant and sustained financial support beginning in 1905.[32] Ellen Scripps was no more interested in the content of the science than the San Diego Marine Biological Association had been. If she gave much thought to the scientists' work, it was to admire the live specimens they displayed in an aquarium-museum (Raitt and Moulton 1967, 35). Ritter was free to design his program without direction or interference from his patrons. In short, the evidence does not support the interest group interpretation.

3.5 Summary

From 1905 until about 1915, William E. Ritter's role was Director of the Scripps Institution. In these early years, the program was coordinated with a single, proto-ecosystemic framework, and with a team of researchers pursuing specialized yet coordinated measurements. In the early years, the Scripps Institution had no connection with, or funding from, the military. During this time, there were no mechanisms whereby these types of factors could influence the core program of research at the Scripps Institution. Although McEwen contracted with electric utilities to provide forecasts for a fee, the incentive was slight and did not shape the core program.

In short, at this time the responsibilities attached to the role of Director were primarily epistemic (research, publish), not political (defend the state) or economic (maximize profits). Ritter, acting in this role position, framed the research program according to his proto-ecosystemic framework. This is an example of the positional fix. Research focused on the marine ecology of the Pacific: coordinated work on plankton and the marine environment. After 1917, when the United States accepted a role in the European war, the Scripps Institution's research agenda took a new and radically different course.

4

Scripps Institution, 1917–1940

Our entry into the war in April, 1917, brought with it a change in attitude toward life, [in] both . . . individuals and organizations. The dominating principle was the winning of the war, to . . . which all personal interests must be subjugated. . . . As one looks back on those hectic years . . . many questions now rise in one's mind as to the ultimate results of these activities.

—Mary Ritter, *More Than Gold in California, 1849–1933* (1933)

The First World War set in motion changes that profoundly altered work at the Scripps Institution. As the United States prepared for war, the state began to build networks for science advice that brought civilian scientists into the war effort. Institutions like the National Research Council's Committee on Pacific Investigations, which William E. Ritter joined in 1917, can be seen as linking some of the responsibilities attached to the role of the United States in the world, a *sovereign role,* to the role of particular scientists, an *epistemic role.* Thereafter, the Director consulted both political and epistemic responsibilities when planning the research program.

The influence of international politics on marine science at Scripps was not, after 1917, monolithic. The United States accepted a role as great power during the war and declined to do so afterwards. To state this more clearly, when the United States entered the First World War, and the Director agreed to serve on the NRC, the Director's role became a political/epistemic hybrid. From 1917 to 1919 the international responsibilities of the United States shaped work at Scripps.

In the interwar years, when the United States shied away from a role as great power, the NRC no longer had specific international political responsibilities. As the U.S. secondary role changed, so did NRC's influence on work at Scripps. Through membership on the NRC committee, the Director viewed marine science in a national perspective and tried to bring the Scripps Institution up to European standards.

This chapter includes two observations on the explanatory variable, social roles: (1) U.S. acceptance of a secondary role (as great power), and (2) postwar U.S. rejection of a secondary role (as great power). It includes two observations on the dependent variable, knowledge: (1) knowledge generated at Scripps during the war; (2) knowledge generated at Scripps in the interwar years (1920s and 1930s). These observations illustrate how institutions shaped knowledge generation at Scripps, through the positional fix.

Section 4.1 provides evidence that changes in the social role (in this case, the U.S. role in the world, mediated by Ritter's position on the NRC) caused changes in knowledge generated at the Scripps Institution. Section 4.2 reviews the history of the interwar years, when the United States retreated from a role as global power.[1] Section 4.3 compares the institutional argument with alternatives, neorealism and the interest group approach.

4.1 Change in U.S. Role (Explanatory Variable)

The United States assumed a role in European international politics when it abandoned its policy of neutrality and entered the war against Germany. Precipitating this event was a German submarine strike at the British passenger ship *Lusitania* in May 1915, which killed more than one hundred Americans on board (May 1959, 135–136). According to the historian John Keegan (1999), the German submarine attacks on the *Lusitania* and other European merchant ships provoked the United States to abandon its stated position of neutrality.

Perhaps more important than the need to protect Allied shipping, in President Wilson's view, was the need to defend the right of Americans to travel abroad. In 1915 and 1916 it became clear to Wilson that if the United States stayed on the sidelines, it would sacrifice an opportunity to shape the postwar order. Stated differently, the President began to expect that a future world order under German domination would be a more hostile place for Americans than an order in which the Allies dominated (May 1959; Sprout and Sprout 1942).

Turning away from his earlier commitment to neutrality, Wilson began to favor war. Opinion in Congress and in the country at large had also turned (May 1959, 433). By entering the war the United States accepted, albeit briefly, a role defending the Allied effort to preserve an international order.

Institutional Networks: Science Advice

As part of the mobilization effort, the United States drew inventors and scientists into military research through a network of advisory committees. These committees began to develop almost immediately after the *Lusitania* incident. In an interview with the *New York Times,* Thomas Edison called on American ingenuity to "overcome" the U-boat threat.

The Secretary of the Navy, Josephus Daniels, after seeing the article, successfully enrolled Edison in a plan to mobilize engineers in the private sector to work on antisubmarine warfare. With Wilson's approval, Daniels tapped Edison to lead the Naval Consulting Board (NCB) (Kevles 1971, 106; Hackmann 1984, 40–41; Keegan 1999, 265). Under Edison's leadership, the NCB linked engineers, largely from industry, with the Navy.

At roughly the same time the NRC, organized under the auspices of the National Academy of Sciences (NAS), began to coordinate university researchers to work on antisubmarine warfare (Kevles 1971,12; Hackmann 1984, 40–41). By institutionalizing science advice, the NRC created a channel through which the military could contract research to universities, and through which researchers could seek federal support for their work. In February 1917 the NRC's first director, George Ellery Hale, focused the NRC on one "critical task": detecting German submarines.

Organized through these networks, most of the early work on hydrophones was conducted at Nahant, Massachusetts, and New London, Connecticut. The Nahant station was jointly authorized by the NRC and the NCB, and brought together engineers from the Submarine Signal Company, General Electric, and Western Electric. The Naval Experiment Station was created in 1917. The United States convened a conference at the New London station in the spring and summer of 1917, bringing British and French researchers together with Americans to focus on problems related to antisubmarine warfare. Through its various com-

Table 4.1
Main American Anti-Submarine Research Centers during World War I

Place	Main Work	Funded by
Nahant, Mass.	Hydrophones; preliminary sea trials	NCB
Naval Experiment Laboratory, New London, Connecticut	Hydrophones; ultrasonics; preliminary sea tests	NRC
Columbia University, New York	Ultrasonics; including amplifiers	NRC
San Pedro Submarine Committee	Quartz, Rochelle salt transducers; magnetostriction	NRC
Schenectady, GEC Laboratories	Rochelle salt; high frequency oscillators; pliotron	NRC assistance
Bureau of Standards	Inspecting and cutting quartz	—
Wesleyan University	Rochelle salt	NRC
Navy Yard, Key West, Florida	Sea trials	US Navy
Pasadena, California	Part of the San Pedro Cttee; cements; power measuring instruments	NRC
WEC Laboratories	Telephonic use of piezoelectricity	NRC

Source: Hackmann 1984, p. 41.

mittees, the NRC contracted specific problems to other research groups around the United States (see table 4.1).

John C. Merriam of the Carnegie Institution in Washington brought Ritter into these circles. Ritter and Merriam had been friends during their student days at University of California, Berkeley, in the late 1880s. Merriam had moved on to assume the Presidency of the Carnegie Institution and was an active member of the American Association for the Advancement of Science (AAAS).

In March 1916, Merriam wrote to Ritter to inform him that the AAAS had created a Pacific Coast Subcommittee, to "encourage and assist" scientific investigation in the Pacific. When the United States entered the war in 1917, Merriam tapped Ritter's expertise in Pacific marine science. By this time, the NAS had created the NRC, which in turn created a Committee on Pacific Investigations. Merriam chaired the Committee, and sent off another letter to Ritter.

War Work, 1917–1919 (Dependent Variable)
In March 1917, Merriam asked Ritter to draw up a plan of wartime research, under NRC auspices, for which the Scripps Institution could be held responsible.[2] Ritter responded that all "organic products of the sea" off the California coast could potentially be involved. Kelp was used for fertilizer, potash, and other chemicals useful to the war effort. Increasing fisheries production would help meet projected increases in wartime demands for food. In 1916 fishing and canning of sardines and tuna in San Diego and San Pedro grossed approximately $4 million per year.[3] As Director, Ritter believed that more research on the location of the fish could help the industry expand.

War work developed as the Director had planned, extending marine biology to serve the state and the industry. The Council of Defense of California funded the additional research, which brought Scripps' research to the service of the National Bureau of Fisheries and the California Fish and Game Commission. In planning the research program, the Director anticipated that fisheries work would yield new information about the abundance of albacore, yellowfin, barracuda, sardines, anchovies, smelt, and mackerel. He tried to bring Scripps' existing research on plankton, the base of the marine food chain, and hydrography, environmental conditions affecting fish, to bear on questions of fisheries abundance.[4]

Longer-Term Effects
When the war ended, so did the flow of funds from federal and state governments for wartime research. Nevertheless, science advisory committees remained in place. These networks continued to shape the research program at the Scripps Institution in the 1920s.

As a member of the NRC, Ritter began to think the core program at the Scripps Institution needed to be reoriented. Like the Kiel school, the Port Erin station, and ICES, the research program integrated physical and chemical observations to provide a picture of an ecological whole. The primary focus was biological, and its focus was on plankton, copepods, and other organisms. During and after the war, Ritter began to change his view of what the Scripps Institution should do.

Evidence suggests it was membership on the NRC that caused him to reorient the Scripps Institution's research program. Membership on the

NRC brought to his attention that the program had not been a good fit with wartime research needs. Since the *Lusitania* incident, the NRC wanted scientists to work on different facets of submarine acoustics, including the development of new instruments, to improve the Allies' ability to detect and to destroy U-boats.

The change can be most clearly conveyed by letting Ritter speak for himself. He wrote about it very plainly in a series of letters in 1922. In August 1922 he wrote to John C. Merriam,

> [The death of a colleague, a marine biologist] brings very vividly into my memory an incident of years ago.
>
> When it was still rather uncertain whether I should ever get fully to my working feet again after my illness in the fall of 1910, one of my chief anxieties was naturally the future of the Scripps Institution. Among the possibilities that occurred to me for making [it] more secure was that of combining it on some basis with the department of marine biology of the Carnegie Institution. . . .
>
> [N]ow as then [my chief concern is the Scripps Institution's] future. . . .
>
> But the problem is quite different now from what it was then. . . .
>
> As you know from our talks on the subject . . . it is now my conviction that the Scripps Institution has made itself into a possible nucleus and ought to become the actual nucleus of a really great American Institution of oceanography, physical and biological.[5]

As he was organizing a meeting in Washington, D.C., at which he hoped to set the Scripps Institution on its new course, Ritter asked Merriam to see to it that a member of the National Research Council would be there. In October 1922, Ritter wrote to Merriam,

> The enclosed letters will inform you of the steps I am taking toward such a conference in Washington [on the Scripps Institution's future] as you suggested in your letter of October 2. . . .
>
> It would seem to be desirable that the Committee on Pacific Investigation of the National Research Council should be represented in this effort.[6]

Ritter's letter to his wife after the meeting in Washington is also revealing:

> I have been thinking very earnestly . . . how I am going to acquaint the staff of the Institution with what has been done; and have decided to write this letter containing the most vital elements and ask you to call the members together at our house and read it to them.
>
> It is of course impossible to go into much detail but I do not think this essential to a pretty clear understanding of what the situation now is.
>
> First as to the conferences that have been held on Institution matters: The luncheon at Merriam's home, President Barrows, Merriam and myself, was the first thing. . . .

> [We reached] agreement that the oceanographic institute in the large way we have been talking for a year and more ought to be the aim. . . .
>
> The second conference, held on the same day, . . . added T. Wayland Vaughan to the group. V. was drawn in because of his previous interest in and wide knowledge of Pacific Investigation matters. He is the virtual chairman of the Committee on Pacific Investigation of the National Research [Council].[7]

In these letters, Ritter reveals the extent to which his participation on the NRC during the First World War changed his vision of what the Scripps Institution could, and should, do. In 1922 the Director thought Scripps should emphasize physical oceanography, not marine biology (as it had since 1905). He thought Scripps should maintain some ongoing connection with the National Research Council, Committee on Pacific Investigations. In this regard, Vaughan's position on the NRC made him particularly attractive as a candidate for the new Director position. Ritter also received Merriam's assurance that the Carnegie Institution would "help" Scripps to develop oceanographic research.[8] These changes, in Ritter's view, would position the Scripps Institution to take a leading role in oceanographic research as applied to the Pacific.

Given his new goal to make Scripps oceanographic, the offer of financial help from the Carnegie Institution must have been attractive. Carnegie had supported work of Vilhelm Bjerkenes and his students in Bergen, Norway, for many years. At that time, physicists in the Bergen school were regarded as the world's most accomplished in the field of dynamic oceanography.[9]

The U.S. responsibilities in the First World War, channeled through science advisory networks, did not merely generate new projects for biologists in La Jolla. These responsibilities caused the Director to rethink the Scripps Institution's basic orientation.[10] In charting its future course, Ritter took care to ensure that his successor could be active on the NRC. He recast Scripps as a center for physical, not biological, oceanography.

4.2 The Interwar Years, 1920–Late 1930s

U.S. Rejects Global Political Role (Explanatory Variable)

The international political responsibilities that had shaped wartime research soon dissolved. After the war, the United States backed away from further involvement in European politics.[11] The Senate refused to ratify a treaty that would have made the United States a member of the League of Nations (Hillman 1952, 422; Gaddis 1997, 34). Carr (1933) wrote, "World leadership was offered, almost by unanimous consent, to the United States . . . [and] it was declined"(34).

Under Presidents Herbert Hoover and Franklin Roosevelt, the United States continued to assert its self-imposed isolation. In 1934 the Congress passed a law banning loans to any government that had defaulted on war debts. The following year the United States declared that a general arms embargo would take effect in the event of war. Congress also prohibited loans to any belligerent power. If (as Hedley Bull argues) the political role of a great power includes the responsibility to prevent a war among other great powers, then the United States clearly sent signals to Great Britain and France that it would not accept such a role (Bull 1977, 207; Kennedy 1988, 329–330).

Martin Wight assesses the U.S. role in the interwar years as follows:

> The U.S. was incomparably the strongest and the most impregnable of the Western powers, but after the First World War she had dissociated herself from them and withdrawn into isolation, repudiating all political commitments outside the American continent and its Pacific outliers. This made it impossible for Britain and France to count on her for support, and easy for their less percipient enemies to suppose they would not enjoy it.[12]

Although the United States had the capabilities to occupy the role of great power, once again it declined to do so.

Mechanisms of Influence: NRC Membership

The U.S. retreat from European politics was felt rather immediately at the Scripps Institution. The flow of funds for war work from the federal government and the State of California dried up by the early 1920s. Research projects on kelp and on fisheries were scaled back and later abandoned.

Other effects of wartime mobilization were more enduring. The University of California changed the organization's name to the Scripps Institution of Oceanography in February 1924, as T. W. Vaughan arrived as the new Director. The institutions of science advice, mechanisms built to mobilize scientists for the war effort, remained in place. These institutional mechanisms continued to shape marine science, despite the political climate of isolation in Washington.

Vaughan was an active member of two NRC committees and the Committee on Oceanography of the National Academy of Sciences. In the late 1920s, this NAS committee assigned itself the task "to consider the share of the U.S. in a worldwide program of oceanic research" (Raitt and Moulton 1973, 108). The committee's chair was Frank R. Lillie, the director of the marine biological laboratory at Woods Hole.

The NAS committee concluded, in a report written by Henry Bigelow, a zoologist at Harvard, that American oceanography had not advanced as far as it should, compared with programs in Europe (Raitt and Moulton 1973, 109; Schlee 1973, 273). The problem was not only that American oceanography did not measure up to European standards but also that research on the Atlantic coast lagged behind research on the Pacific coast.

Guided by the NAS report, the Rockefeller Foundation made grants to a number of programs in marine science. The grants made it possible for Lillie, Bigelow, and others to establish the Woods Hole Oceanographic Institution in 1930 (Schlee 1973, 275). Rockefeller money also went to the Bermuda biological station and to new laboratories at the University of Washington. The Foundation concluded that the Scripps Institution was in good shape compared to research facilities on the East Coast, and made a comparatively small grant to Scripps for a new laboratory.

Advisory networks under NAS auspices, in short, were important in that they assigned to men like Vaughan, Lillie, and Bigelow the responsibility of thinking about oceanography in a national perspective. In other words, oceanographers were not simply mapping out problems for research in relation to the present state of the field. Under NAS auspices, they were also comparing American oceanography to the state of the field in other countries. As historian Susan Schlee (1973) put it, the NAS wanted to know, "Are American oceanographers keeping up with the Petterssons, the Murrays, the Hjorts?" (273). By stimulating a series of grants from the Rockefeller Foundation, the NAS precipitated the largest expansion of American oceanography prior to 1941.

The Director's Role

Work with the NAS was just one part of Vaughan's role as Director. A major part of his job, as Vaughan saw it, was to transform what had been a biological station into an oceanographic institution.

One important task was to assemble a staff with expertise in marine science. When he arrived in La Jolla in 1924, Vaughan announced his intention to transfer Francis Sumner, whose research consisted of experiments on deer mice, to the University of California at Berkeley. Vaughan thought Sumner's work did not fit within the Scripps Institution's new mandate. Sumner was reluctant to change direction.[13] But rather than move to Berkeley, Sumner shut down the "mouse house" and turned his attention to the physiology of fishes.

The Director added new staff as positions became available. Erik K. Moberg, a specialist in marine chemistry, arrived in 1925. Vaughan hired Claude Zobell, a microbiologist, to fill a line opened by C. O. Esterly's death in 1928. Dennis L. Fox, a specialist in the physiology of marine organisms, arrived in 1931.

As Director, Vaughan sought to improve the flow of oceanographic data to the research team. In 1925, the Scripps Institution acquired a new vessel, named the *Scripps*, previously used as a fishing vessel. Although the University of California was gradually increasing the amount of money available for salaries, staff assistants, and equipment, Vaughan was unable to add enough new research vessels to gather the quantity of data the staff needed. Instead, he sought out partnerships with other government agencies.

Vaughan described a number of these partnerships in a letter to U.C. President Campbell in October 1924. An arrangement with the U.S. Coast and Geodetic Survey put at Scripps' disposal, year round, two seagoing vessels. The U.S. Bureau of Lighthouses and the Hopkins Marine Laboratory collected water samples and temperature records at a number of sites along the Pacific coast. The U.S. Bureau of Soils and the U.S. National Museum sent samples of marine sediments. Naval ships and lighthouses took water samples and hydrographic measurements. These data-gathering arrangements improved the quantity of data available for the research team.

Knowledge Generated (Dependent Variable)

The partnerships with the U.S. Coast and Geodetic Survey, the Bureau of Lighthouses, and other agencies improved the quantity of information available to researchers. In a 1931 report, Vaughan noted that observations of ocean temperatures increased from 2,170 in 1923 to 19,444 in 1925, and salinity readings from 1,705 in 1923 to 5,573 in 1925. W. E. Allen used some of these data to publish a study on the abundance of plankton relative to the hydrographic conditions in the area (Shor 1978, 230).

One problem with data gathered by people outside the Institution, however, was that the data were primarily collected from areas between major ports. To gather samples in a more meaningful way would require a fleet of research vessels. The U.S. Navy was not funding this type of research in the 1920s and 1930s. E. W. Scripps' son Robert, who had promised to find Vaughan a new research vessel, was unable to do so after his fortunes declined with the 1929 stock market crash. Without adequate financial support, Vaughan found it impossible to acquire an adequate number of research vessels.

Another problem, in Vaughan's eyes, was that the core of the Institution's oceanographic program should have been physical oceanography. However, by the mid-1930s the program was still quite limited in this area. McEwen continued his studies of ocean temperatures and climate, which until 1932 had the financial support of local power companies. McEwen also supervised a project to compile oceanographic observations into a database covering the period 1904–1934. These data were transcribed from coding sheets to punch cards, then sent to the Navy Hydrographic Office in Washington for processing. The cards were eventually returned to Scripps for storage. The Institution's physical oceanographic work was limited to this kind of data gathering and tabulation (Shor 1978, 242–243).

Research developed along a number of individualized lines. Sumner began work on how marine organisms change color. A new recruit, Dennis Fox, worked jointly with Sumner on a number of these projects (Shor 1978, 215). Claude Zobell studied the concentration of organic compounds on solids submerged in sea water (Shor 1978, 225). Moberg initiated a number of studies in marine chemistry, including a manuscript Vaughan considered to be quite important, on the effects of acids in sea water on concentrations of calcium carbonate (Shor 1978, 323). Elements were in place for coordinated work, but the missing piece of the puzzle remained an expert in dynamic, physical oceanography.

Vaughan's Search for a New Director

As Vaughan retired, he looked to the Bergen school for a possible replacement. He had been building ties with his Norwegian colleagues for a number of years. Vaughan had met Harald Sverdrup, a young oceanographer from the Bergen school, in the winter of 1921-22. At that time, Sverdrup was visiting at the Carnegie Institution in Washington (Day 1999, 72).

In March 1926, Vaughan wrote to Vilhelm Bjerknes, then head of the Geophysical Institute. Vaughan asked Bjerknes who, in his opinion, were the best minds in the field. Bjerknes replied that in his view, V. Walfrid Ekman, Fridtjof Nansen, and Bjørn Helland-Hansen were the "most eminent workers in dynamical oceanography."[14] Sverdrup was a student of Helland-Hansen.

By that time, Vaughan was actively seeking his replacement as Director. He wrote to U.C. President Sproul,

> I do not know whether you are acquainted with Helland-Hansen's work. He is the director of the Geophysical Institute at Bergen, Norway, which is rather surely the most important institution for the study of the problems of dynamical oceanography. For some years Professor V. Bjerknes was associated with the Institute, when Helland-Hansen was one of his juniors. Now Helland-Hansen is the director and has associated with him H. U. Sverdrup. The National Academy medal for achievement in oceanographic research was awarded to Helland-Hansen at the meeting of the Academy last April.[15]

Having broached the topic with Sproul, Vaughan went after Sverdrup. On the Director's behalf, Helland-Hansen asked Sverdrup, who was then working at the Geophysical Institute in Bergen, whether he would be willing to consider the position as the Scripps Institution's next Director.

Sverdrup had planned to take over as director of the Geophysical Institute upon Helland-Hansen's retirement. He accepted the job in La Jolla with the caveat that the position would be temporary.[16] Sverdrup arranged for a three-year leave of absence from Bergen and arrived at Scripps in September 1936.

In recruiting Sverdrup, Vaughan brought to Scripps one of the world's leading dynamic oceanographers. Sverdrup had studied with Vilhelm Bjerknes at Leipzig University. In 1917 he was the lead scientist on the Norwegian North Polar Expedition (the MAUD Expedition), a seven-year adventure in which he amassed oceanographic data. Publications from the expedition sealed Sverdrup's reputation as the world's leading expert on oceanic-atmospheric interactions.

Sverdrup became a professor at the Geophysical Institute in Bergen in 1931, as well as the first research professor at the new Christian Michelson Institute. Part of that year Sverdup expanded his contacts at the Carnegie Institution in Washington (Shor 1978, 10; Oreskes and Rainger 2000). In the early 1930s, Sverdrup participated in two Norwegian government–sponsored expeditions, the Wilkins-Ellsworth Arctic Expedition on the submarine *Nautilus,* and an expedition to Spitsbergen (Svarlbard), where he studied heat transfer between the atmosphere and the snow (Shor 1978, 9–10; Oreskes and Rainger 2000, 309). With Sverdrup as director, Scripps was well positioned to take the lead in research on dynamic oceanography in the Pacific.

Research at Sea: 1936–1940 (Dependent Variable)

When he took over as Director, Sverdrup was best known for oceanographic research expeditions to the polar north. But soon after he arrived in California, the Scripps Institution's sole research vessel caught fire and sank. Undaunted, Sverdrup acquired and outfitted a new vessel, the *E. W. Scripps,* with financial help from Robert Scripps and from the University (Friedman 1994, 31). His ambition was to pry researchers out of their laboratories and to get them out to sea.

In general, Sverdrup was not impressed with the quality of scientific work he found at the Scripps Institution. He thought the data Vaughan had secured through cooperative arrangements were "completely worthless" (Friedman 1994, 29). Further, Sverdrup complained, "With all due respect to McEwen as a mathematician: McEwen is not a geophysicist. He has no sense for observations . . . [and the institution has] no other oceanographer."[17] To Sverdrup, oceanography meant physical oceanography, and he did not find any such thing in La Jolla.

In Sverdrup's view, the Institution lacked a coordinated program of research. Vaughan had done his best to assemble a team of marine scientists, but he himself was not capable of designing a coordinated research program in physical oceanography. And, as Sverdrup pointed out, the Institution didn't really have a physical oceanographer. According to

Sverdrup, the Institution needed a coordinated research plan for the study of Pacific ecosystems (Friedman 1994, 32).

Part of Sverdrup's plan was to survey the ecosystems along the California coast. Ecological studies of the California Current began in the spring of 1937. Sverdrup organized three cruises aboard the *Bluefin,* a vessel belonging to the California Department of Fish and Game. Work on the California Current provided important insights into coastal upwelling, including a relationship between physical and chemical oceanographic factors and areas of maximum spawning of sardines (Scheiber 1988; Friedman 1994, 29; Nierenberg 1996, 13).

Sverdrup's plan also included two expeditions to the Gulf of California on the *E. W. Scripps.* The 1939 expedition, which Sverdrup led, was the first comprehensive hydrographic survey of the Gulf of California. The 1940 cruise, led by Charles Anderson, Roger Revelle, and Francis Shepard, concentrated on land and submarine geology. Sverdrup had organized a coordinated research program, which was guided by his theoretical and practical expertise as a physical oceanographer. This was roughly what researchers at Scripps were doing when the program was upset, once again, by war.

4.3 Alternative Explanations

This chapter has argued that the positional fix—the use of the Director's role position—shaped and reshaped research at the Scripps Institution. When the Director became a member of the NRC, both political and epistemic responsibilities framed the research program. As the U.S. role in the world changed, mediated through the Director's position on the NRC, the research program changed as well.

By contrast, according to neorealism, the state would fund those aspects of marine science that furthered its military interests. In this model, interests are derived from a state's position in the system (Waltz 1979). If the United States had the material capabilities to act as a great power, it would fund military applications of marine science to increase its power (or security). The evidence does *not* support a neorealist interpretation.

Judging from military capabilities alone, the United States continued to be one of the world's great powers in the years between the two World

Wars. Therefore, according to neorealism, the military should have continued to fund military applications of oceanography in the 1920s and 1930s. The Scripps Institution's war work should have been funded at a steady rate. The evidence does not support this interpretation. After World War I, the military discontinued funding for military applications at the Scripps Institution.

Nor does the evidence support the interest group explanation. There was as yet little commercial interest in marine science. What little work McEwen did for power companies was discontinued under Sverdrup. Fishing firms did not yet approach marine scientists for help with declining yields: that came later. In short, the interest group approach cannot explain changes in the research program during these interwar years.

4.4 Summary

As Mary Ritter observed, the U.S. "entry into the war in April 1917 brought with it a change in attitude toward life, both as to individuals and organizations." During this period, 1917–late 1930s, the U.S. role changed twice. First, the United States accepted a role by entering the European war (explanatory variable). During the war years, the rights and responsibilities attached to the role of great power shaped research at the Scripps Institution. Plankton research, previously the core of the Institution's program, was set aside. War work, primarily research into fisheries and kelp for war supplies, supplanted it (dependent variable).

Mediating between the U.S. role and the role of Scripps Director were networks of science advice, like the NRC. When Ritter became a member of the NRC, his role changed. Previously, his responsibilities had been primarily epistemic: research, teach, publish, and organize the research, staff, and resources to carry out a proto-ecosystemic research program in marine biology. After he joined the NRC, the rights and responsibilities the United States accepted as a great power were added to his role as Director. These included, for example, the responsibility to assist European allies in the First World War. As a result, the Director reshaped the Institution's research program, both for the short term, by conducting war research, and for the long term, by reorienting the Institution from biology to oceanography.

The U.S. role changed again after the war, from great-power status to isolation in the 1920s and 1930s (explanatory variable). When the United States rejected a continuing role as great power, funding for military applications of marine science dried up. Though the funding dried up, the institutional mechanism that brought the U.S. political role to bear on the research conducted at Scripps—the Director's membership on the NRC—stayed in place. The Scripps Institution continued to transform itself from a marine biological station to an oceanographic institution. The research program changed from a focus on plankton ecology to a focus on coordinated research in physical, chemical, and biological oceanography (dependent variable).

Mediating the connection between the U.S. role and the Director's role was the NRC. In the interwar years, the NRC continued to shape research, although the United States did not attach specific defense-related rights or responsibilities to committee membership. Members of the NRC focused on how U.S. national oceanography compared with that in other countries, especially in Europe. In other words, membership on the NRC continued to influence the ways in which the Director of the Scripps Institution planned research, even though the U.S. government ceased to fund military applications of oceanography. Vaughan, an active member of the NRC, attempted to build the Scripps Institution's program in physical oceanography, with an eye to work at the Bergen school.

By the late 1930s, the Scripps Institution was poised to become one of the world's leading oceanographic institutions. Its new director, Harald Sverdrup, was an accomplished physical oceanographer. Not since Ritter's time had the scientific staff undertaken coordinated work at sea. The Scripps Institution was pursuing physical, hydrographic, and biological surveys of the California Current and the Gulf of California.

But as the Second World War began, Sverdrup's research agenda unraveled. After Germany invaded Norway, he gave up his plan to return home.[18] When the United States mobilized for war in 1941, it once again assumed a role as a great power. Its international responsibilities, now reactivated, were once again channeled through science advisory networks. Sverdrup, as Director, saw the research agenda he had developed for the Scripps Institution swept aside.

5

Scripps Institution, World War II, and the Cold War

The general research in the marine sciences at the Scripps Institution has been greatly curtailed in 1942–44 because work at sea was discontinued in July, 1941, because staff members are on war leave or have leave of absence to conduct war research at other establishments, and because more and more war research has been taken up at the Scripps Institution.

—Harald Sverdrup (as Director of the Scripps Institution), April 1944, report to the University of California[1]

During and after the Second World War, the U.S. embraced a new global role, as a great power (or hegemon) responsible for maintaining a global order.[2] As the Cold War developed, a core part of this global role was to maintain a deterrent force, a stockpile of atomic and nuclear weapons intended to deter a Soviet first strike.[3] The Scripps Institution became heavily involved with carrying out oceanographic research connected with a series of atomic bomb tests, from 1946 until these tests were curtailed in the early 1960s (Seaborg 1983, 282).

This chapter continues to illustrate the positional fix: the U.S. global political role, mediated through the Director's role and various advisory committees, profoundly reshaped the postwar research program at the Scripps Institution. Section 5.1 documents how the new U.S. role (hegemon) began to shape research at Scripps. Networks of science advice brought the domains of national security and marine science together. Section 5.2 explains how, in the Cold War era, Scripps became further enmeshed in military-epistemic networks. Through participation in these networks, Directors acquired political responsibilities, which influenced research practices.

Sections 5.1 and 5.2 can be seen as two sets of observations on the explanatory and dependent variables. First, during the war, the United

States accepted a role as a great power (explanatory variable). Oceanographic research changed to focus on military applications (dependent variable). Second, after the war, the United States accepted a role as hegemon (explanatory variable). As in previous decades, networks including, but not limited to, the NRC linked the U.S. role (hegemon) to the role of Director of the Scripps Institution. During the Cold War era, the Director translated the U.S. global responsibilities into a research agenda, expanding traditional lines like undersea acoustics and adding new ones, including thermonuclear tests, pioneering work on climate change, and radiation ecology (dependent variable). Viewed in historical perspective, the Second World War and the Cold War magnified the extent to which changes in global political institutions reshaped the research program at the Scripps Institution. Section 5.3 reviews alternative explanations.

5.1 Global Political Institutions: World War II

Near the end of 1941, the United States entered the Second World War. According to Bull (1977), as it entered the war, the United States once again accepted a role as a great power. The network of science advisory committees that proliferated during and after the war can be seen as channels through which people translated the responsibilities attached to the U.S. secondary role into a research agenda. Scientists participating in national security networks helped to frame problems that were of interest to the military, and to translate scientific findings into military terms.

Mechanisms of Influence

According to Kevles (1971), a web of advisory committees grew rapidly in the Cold War era. Before the war, the core of the interface between academic researchers and the military had been the National Defense Research Committee (NDRC). Under NDRC auspices, a division was created (Division 6) to specialize in underwater sound (Kevles 1971; Hackmann 1984; Mukerji 1989). The Navy created two university-affiliated laboratories: the Columbia University Division of War Research (CUDWR) in New London, Connecticut; and the University of California Division of War Research (UCDWR) at Point Loma, near San Diego (Hackmann 1986, 104). These laboratories were devoted almost exclusively to submarine and antisubmarine warfare (Revelle 1969, 8).

Division 6 began to operate under the umbrella of the Office of Scientific Research and Development (OSRD) in mid-1941. OSRD assembled a staff of about 1,500 people and supported about 15,000 private contractors throughout the United States (MacLeod 2000, 13). The former Vice President of MIT and President of the Carnegie Institution in Washington, D.C., Vannevar Bush, was OSRD's first chairman (Kevles 1971; Mukerji 1989; MacLeod 2000). Vannevar Bush advised President Truman directly.

The Navy used these hybrid networks to contract research out to a number of laboratories, including its Division 6 labs, CUDWR, and UCDWR, but also to the Harvard Underwater Sound Laboratory, Bell Telephone, General Electric, Gulf, and the Woods Hole Oceanographic Institution (WHOI) (Lasky 1975, 886; Hackmann 1986, 104).

Roger Revelle, Naval Oceanographer Roger Revelle, who had received his Ph.D. degree from Scripps in 1936, was part of this interface between the university and the military. By the end of 1942 the Navy had put him in charge of a Bureau of Ships subsection to study sonar design. He became a project officer for the UCDWR, WHOI, and the Navy's underwater sound laboratory in New Haven, Connecticut. Revelle was part of a small group that planned oceanographic research and contracted it out to Division 6 labs and other organizations.

An official with the Navy summarized Revelle's work at the Bureau of Ships in a 1948 letter to Carl Eckert (then the Scripps Institution's Director):

> [In 1944 Revelle's] duties in the Bureau of Ships were enlarged to include the technical planning and guidance of all oceanographic and related research under the cognizance of the Bureau. These involved research on ocean waves and surf, the effects of surf and beach conditions on the performance of landing craft . . . and [the effects of] all other oceanographic factors on submarine operations. All of these projects yielded important results which were brought to practical use during the later part of the war. For example, the Bureau of Ships research on life rafts formed the basis of search procedures for air sea rescue through the Pacific Theatre. The research on waves and surf led to the development of surf forecasting methods for amphibious operations and to means for estimating the percentage of casualties to a landing craft under different surf conditions. . . . *[Revelle] was to a considerable degree responsible for the formulation of the research projects; for stimulating scientists to undertake the work; for guiding their work towards problems of Naval importance; and for translating the results*

obtained into Naval terms. It is no exaggeration to say that the large role which oceanography now occupies in the Navy research program is in part due to Dr. Revelle's effectiveness and foresight in planning and promoting the Bureau of Ships research during the war [emphasis added].[4]

Revelle, at the Bureau of Ships, occupied a hybrid role. He had epistemic responsibilities, in that he was supposed to organize other scientists to do certain kinds of research. At the same time, he had responsibilities that derived from the U.S. role in the world. Attached to Revelle's job in the Bureau of Ships, albeit separated by at least three degrees, was the responsibility to think about deterrence. President Truman, as Commander in Chief, had the primary responsibility. The President assigned certain tasks to the head of OSRD [Vannevar Bush], who oversaw the head of OSRD's Division 6 [Lyman Spitzer, Jr.], who tapped the head of the Bureau of Ships subsection on Sonar Design [Revelle]. Revelle's role in the Bureau of Ships entailed both epistemic and political responsibilities.

What Revelle did in this role can be seen as two-way translation. He translated the responsibility of deterrence into the language of science. He defined areas of research, even in areas where the Navy did not have a clear idea what actions would be consistent with its goal of winning an atomic war. For example, in 1946, naval officers did not know what the effects of an underwater explosion would be on ships, on the marine environment, on fish, or even on the climate. Given the goal of winning a war (with the USSR), it became necessary to fix certain beliefs about what atomic weapons would do if detonated at sea. In 1946 no one knew. Would there be tidal waves? Would radioactive fallout be contained locally, be diffused globally? Would warships withstand the blast? The Navy needed to establish certain beliefs before it could decide whether it would be rational to prepare to use atomic weapons in naval combat.

In more traditional areas, like submarine warfare, naval strategy was more clearly defined. Revelle could develop new lines of research to enable the Navy to improve existing operations. For example, he could explain to the Navy why it needed to learn about the properties of low-frequency sound as it moved in water. He could explain why the Navy needed more detailed and comprehensive maps of the sea floor. Revelle's position at the Bureau of Ships was hybrid, in the sense that he was acting simultaneously as an oceanographer and as a "Navy man."

Harald Sverdrup, Scripps' Director during the War Unlike Revelle, Sverdrup found it impossible to maintain stable contacts with the military (Nierenberg 1996, 16). The Navy appointed Sverdrup to run UCDWR in the summer of 1941. But the following spring, the United States lifted his security clearance and pulled Sverdrup from the project.

Sverdrup himself did not seem to know why. A report to the University of California states that his "services were discontinued . . . because he was not permitted to work with confidential matters, not yet being an American citizen."[5] In a letter dated June 15, 1942, to U.C. President Sproul, Sverdrup put it this way: "Now regarding my own personal situation. You may have learned that at the beginning of March, I had to leave the Project at Point Loma because the Navy regulations do not permit a person with relatives in an occupied country to work with classified material."[6]

Archival evidence suggests that the FBI believed Sverdrup to be a Nazi sympathizer. Some of the charges on which the FBI based its report were anonymous. Others apparently came from staff members at Scripps, who were unhappy with Sverdrup's leadership.[7] Having been denied security clearance, Sverdrup was limited to supervising a small oceanographic division at Point Loma (Oreskes and Rainger 2000, 334).

Sverdrup maintained contact with Revelle at the Bureau of Ships. In late 1942, Revelle was appointed to the Joint Chiefs of Staff Committee on Meteorology (JMC) (Morgan and Morgan 1996, 30–31; Oreskes and Rainger 2000). As a member of the JMC, Revelle convinced the Navy to appoint Sverdrup to lead a research project on sea swell and surf. Apparently as a result of Revelle's aggressive defense of Sverdrup's reputation, the FBI issued a new, favorable report on Sverdrup's character in March 1942 (Oreskes and Rainger 2000, 345). Thereafter, Sverdrup and his younger colleague Walter Munk developed surf forecasting methods that the Allies used to plan amphibious landings (Morgan and Morgan 1996, 31).

Throughout 1942, Revelle and Sverdrup remained in contact. Sverdrup sent updates on his and Munk's work, and requested details about project budgets and administrative matters."[8] As if to clarify that he was writing qua an individual, at the top of each letter to Revelle, Sverdrup typed PERSONAL. In one letter, he observed, "I still have to write you personal letters because I do not know what official relations I have with the Navy."[9] He wrote personal letters because, as Director, he was excluded from military projects.

During the war, the oceanographic projects that mattered most in intellectual and practical terms were defense-related. This created a dilemma for an individual like Sverdrup, who was allowed to occupy only an epistemic role, a role as scientist. On the one hand, Sverdrup was regarded as one of the world's most accomplished experts in dynamic oceanography, and he wanted to continue as such. On the other hand, perceived as a foreigner, he operated under a cloud of suspicion. He was shut out of the most active projects in physical oceanography, which required security clearance. A more serious problem for Sverdrup was that, in the years immediately after the war, the ability to fit into scientific advisory networks became an even more important part of the Director's role.

To do physical oceanography during the war virtually required a scientist to occupy a hybrid role. Not completely so: Sverdrup and Munk did important work on surf and swell. But much of the important work in submarine warfare was classified, and questionable persons were excluded (Nierenberg 1996, 16). Sverdrup's experience provides a rather graphic illustration of how political institutions shaped marine science during the war. As an individual scientist, in principle, Sverdrup could pursue whatever line of research interested him. But as Director of the Scripps Institution, he would need to know what people were doing, and lacking security clearance, he was denied access to this knowledge. What is more, in order to direct the best scientists on his own staff, who had taken "war leave" to work at Point Loma, Sverdrup needed clearance. Because the United States did not allow him to occupy a hybrid position, Sverdrup's work as an oceanographer was constrained. His effectiveness as Director was diminished, particularly compared to what it had been before the war.

Knowledge Generated (Dependent Variable)

The war upended established lines of research and set new ones in motion. The research cruises along the California coast and the Gulf of California were terminated by the middle of 1941. By the spring of 1942, Sverdrup had obtained a limited clearance to work on meteorological projects. Sverdrup and Munk developed a set of dynamic equations to predict the length, height, and speed of waves under different weather conditions, and as they broke upon reaching the shore. The United States used these techniques when planning operations off North Africa, Sicily, and Normandy. Revelle later recalled, "[In Sicily and Normandy] the German commanders were apparently confident that the surf would be too high for successful landings. But the forecasters were able to predict that the landings would in fact be possible, and this proved to be so" (Revelle 1969, 11).

The Navy also found problems it needed biologists to help solve. UCDWR tapped Martin Johnson to investigate the source of crackling sounds that disabled sonar. Johnson discovered that the sound was made by snapping shrimp. Since the noises would occur only in the areas habitable for shrimp, Johnson was able to predict when and where these crackling noises should occur, throughout the Atlantic and Pacific.[10] This removed one obstacle to effective use of sonar. After the war ended, the Navy and the university found ways to continue this kind of hybrid research.

5.2 U.S. Role as Hegemon (Explanatory Variable): The Cold War

The United States and the USSR emerged from the Second World War as rival hegemons, each assuming a role with respect to a network of satellite states. The postwar order can be seen as institutional in the sense that both countries tacitly accepted roles in maintaining it (Kennedy 1988, 365; Gaddis 1997, 12–23). Each accepted a number of rights and obligations to modify its policies in light of hegemonic responsibilities.[11] As part of its secondary role, the U.S. developed and maintained an arsenal of atomic, and later thermonuclear, weapons.[12]

The U.S. role in the world profoundly affected scientific research. After the Second World War, Western Europeans expected, and Americans expected, that the United States would build enough weapons of mass destruction to deter a Soviet first strike. Physicists around the country, including oceanographers at Scripps, were drawn in (Kevles 1971). The military established proving grounds in New Mexico, and on Bikini and Enewetok atolls in the Pacific Ocean.

As applied to the Pacific, the U.S. role entailed dominance of the entire ocean area, above and below the surface. The comprehensiveness of this responsibility was completely new. After the war with Spain, American naval planners had been concerned with preventing European powers from taking military action in the Western Hemisphere (Sprout and Sprout 1942; May 1959). The global role, by contrast, generated a new way of looking at the Pacific. In addition to thinking about the ocean as a proving ground for atomic and nuclear weapons, naval planners began to imagine it as a potential battleground. In order to deter a possible Soviet underwater attack, the Navy wanted to dominate not only the surface of the Pacific but the entire basin.

Channels of Institutional Influence

During the early Cold War era, the United States continued to organize research through a number of institutional channels. Resources continued to flow to military projects, including work in oceanography. Immediately after the war, still enlisted, Revelle helped to organize networks linking the Navy to the University of California. The Office of Naval Research (ONR) assumed the wartime activities of OSRD. Revelle was put in charge of ONR's Geophysics Branch. His job was to

> administer, coordinate, and direct research in the field of geophysics, principally oceanography, meteorology, and geology; as applied to scientific warfare . . . [to] evaluate reports as received from research institutions and contractors and . . . [to] advise cognizant Bureaus and Offices of the Navy Department of their possible application in the naval organization.[13]

In this position, Revelle was part of the core group that advised the Navy as to how oceanographic research could be used to develop military strategies, given the overall political goal (or responsibility) of deterrence.

Working with Lyman Spitzer, then head of the Navy's submarine research division, Revelle drafted a letter to U.C. President Sproul. The contract formalized an estimated $175,000 per year in research on underwater sound to be conducted at the Scripps Institution. This was a sizable amount, compared to the total annual research budget of the University of California, and compared to the total annual budget of the Scripps Institution, both about $100,000 at that time.

Sproul agreed to create a new administrative unit, called the Marine Physical Laboratory (MPL), in San Diego, to handle ONR contracts. MPL was the peacetime successor to the UCDWR. MPL was established as a center for basic research on oceanography as applied to antisubmarine warfare.[14] Reflecting on the construction of these hybrid institutions, Revelle recalled, scientists . . . had helped administer the application of scientific and technological knowledge to the problems of the war. *We were all filled with missionary zeal to build up fundamental scientific research in the United States to the same level of effort and government support as had been given . . . during the previous five years.* [emphasis added]

By making Navy support for oceanography a long-term commitment, Revelle sought to make hybrid positions, like the one he had occupied in the Bureau of Ships, a permanent part of the establishment. The contracts that Revelle (at ONR) established with the University of California were closer than arms-length agreements. Revelle, a scientific insider, was drafting them from within the Bureau of Ships. A short time later, Revelle, as director of Scripps, was seen as a "Navy man." The postwar contracts between the university and the Navy were structural because they were long-term in nature. As such, they bound the university and the Navy together in a set of hybrid institutions.

Before leaving the Navy's Bureau of Ships, Revelle was assigned to Joint Task Force One (JTFO), a joint Army-Navy unit that conducted the first postwar atomic tests on Bikini atoll. Revelle led the oceanographic and geophysical research under JTFO auspices. According to John Isaacs, part of the survey team, "The Crossroads scientific program was Revelle's idea, and he organized it single-handed. Basic scientific understanding of many of the effects of atomic weapons still rests on this one, truly scientific operation."[15] Crossroads provides a graphic example of the kind of research Revelle organized as an oceanographer in the Navy. As a scientist, he designed a research program. As a Navy man, he directed the team as its Commander. Revelle translated the U.S. responsibility for deterrence (attached, albeit three steps removed, to his role) into an oceanographic research program. He translated research results into terms the Navy used to decide whether it should plan to wage atomic war at sea.

Knowledge Generated: Crossroads (Dependent Variable)

Histories of the scientific work attached to Crossroads tend to credit the Navy for a broad interpretation of what kinds of projects counted as relevant. Revelle, who orchestrated the oceanographic program, seemed not to mind that his scientific colleagues could find ways to attach projects they thought of as "pure science." He encouraged it, and he himself did it. ONR's liberal approach to funding research continued throughout the 1950s and 1960s.

JTFO assigned to Revelle the responsibility to study the diffusion of radioactive wastes in the ocean, and the effects of the bomb on the marine environment. Acting as a naval officer but tapping his scientific colleagues, Revelle pulled together a team of researchers from the U.S. Geological Survey, the Fish and Wildlife Service, the Scripps Institution, the Woods Hole Oceanographic Institution, and the Universities of Michigan, Southern California, Washington, and California. Among the members of the Crossroads oceanographic team were Munk, Johnson, and Isaacs (then a research engineer at Berkeley).

JTFO intended Crossroads to test the effect of atomic explosions on ships. The team detonated two devices, code named Able and Baker. Able exploded approximately 500 feet above the lagoon's surface, and Baker several hundred feet below it. Research focused on three main problems. The team studied the waves created by the atomic explosions, the circulation of radioactive materials in the lagoon, and the effects of the blast on living organisms, including coral.

Baker exploded beneath the lagoon's surface on July 25, 1946. In preparation for the test, Revelle's team had experimented with TNT in the Chesapeake Bay. Based on the results, researchers estimated the height of waves a 20-kiloton device might generate. To measure the height of waves, researchers erected towers in Bikini lagoon equipped with cameras, and took time sequence photographs. They also erected poles around the edge of the lagoon, and attached small cans to the poles at 1-foot intervals. Judging from the photos, and the height at which they retrieved cans filled with water, researchers estimated the wave height at approximately 94 feet. This was roughly in line with expectations (Weisgall 1994, 244). However, the oceanographers did not anticipate the column of spray, or base surge, that lifted a mile in the air and collapsed in a doughnut-shaped mass back into the lagoon (Revelle 1969, 4). As it collapsed, the base surge inundated men and ships with radioactive water and debris.

In the months before Crossroads, researchers also conducted a general survey of the ecology of the Bikini atoll. In 1947, Revelle led another team of oceanographers to resurvey the site. The team produced a general study of the biological effects of radiation Baker had released into the lagoon.

Another problem was to assess the effects of the blast and of radiation on the ecology and geology of the atoll. Munk used Crossroads to generate two papers. In "The Circulation of the Bikini and Rongelap Lagoons," Munk reports his discovery of a new circulation pattern. The system consists of two "counter rotating compartments which move in a clockwise sense in the southern portions and in a counterclockwise sense in the northern portions."[16] On the other hand, in a report advising the government about radiation in the lagoon, Munk wrote, "The slowness with which the lagoon is flushed means that only small amounts of contaminated water enter the ocean at any time—direct outside contamination by rain storms will be a more important factor, and the efforts in contamination should be directed toward this end."[17] According to Mukerji (1989), Munk saw the first paper as "pure science" and the other as science advice to the government.

Revelle used Baker to resolve a longstanding dispute in biology and oceanography. Between 1892 and 1902, Alexander Agassiz had undertaken a number of research expeditions to coral-bearing areas of the ocean, out of dissatisfaction with Charles Darwin's theory of reef formation. Darwin's idea was that coral reefs form atop extinct volcanoes, as they gradually sink into the sea floor (Deacon 1980, 109; Rainger 2000, 367–368). To test Darwin's theory, the team drilled 800 meters into the atoll (Day 1985). Revelle published the results as a piece of "pure research," confirming that Darwin's hypothesis explained the formation of the Bikini atoll.

Crossroads illustrates the sense in which researchers had hybrid roles. They generated research that would help the Navy to develop its strategic plans. The oceanographic work at Crossroads led to a research program that ultimately raised red flags about the adverse effects of atomic and nuclear weapons on oceanography and fisheries (NAS/NRC 1957).

Resistance to Revelle's Appointment as Scripps' Director In early 1947, Sverdrup announced that he planned to retire and return to Norway.[18] He made it known to the faculty that Revelle was his choice as the next director. Sverdrup had been impressed with the extent to which Revelle had persuaded the Navy to support oceanographic research. Perhaps more important, Revelle had considerable experience with research at sea. Sverdrup considered it essential for Scripps to get its oceanographers out of laboratory buildings and onto oceangoing vessels. Revelle agreed to return as associate director under Carl Eckert, then in charge of Marine Physical Laboratory.[19] The university appointed Eckert as Director in March 1948, with the tacit understanding the appointment would be temporary.

Revelle returned to the Scripps Institution in 1948, with an appointment as associate director and professor of oceanography. He remained a civilian consultant to ONR. Controversy raged over whether Revelle, who had the support of Sverdrup, Munk, and Eckert, would take over as Director. Revelle faced a "fireball of opposition" (Munk 1991, 9).

Part of the problem was Revelle's administrative habits, which even he admitted had been less than adequate.[20] According to Munk (1991), another source of opposition to Revelle's appointment "was a majority of Scripps faculty . . . [which] held that it was about time to have a biologist for director" (9). Revelle recalled, "[A faculty member opposing him] thought we ought to have a biologist as director, not a military man, not a Naval man, of all things not a Naval man!"[21] In July 1951, the university appointed Revelle to the Directorship (Day 1985, 14–15).

It would have been difficult for the university to overlook the sparkle of Revelle's candidacy, in the light of the Navy's recent largesse. In the late 1930s and early 1940s, the Scripps Institution's total annual budget was less than $100,000.[22] This was roughly comparable to that of Woods Hole and more than double that of the Hopkins Marine Laboratory and the University of Washington.[23] In February 1945, the Navy Hydrographic Office awarded Scripps a single contract worth $140,000 for oceanographic work at Point Loma.[24] In early 1946, the Navy's Bureau of Ships sent a letter to Sproul, estimating future naval support of $175,000 per year for antisubmarine warfare. This was the Navy's first long-term commitment to university-based research (Morgan and Morgan 1996, 33). At the same time, the UCDWR's operations were transferred to Scripps' new Marine Physical Laboratory on the La Jolla campus. By 1948, the Scripps annual budget was close to $1 million (40).

The Importance of Advisory Committees In an interview in 1985, the historian Sarah Sharp asked Revelle how he saw his role as Director. Revelle's reply lays bare not only the importance of advisory committees but also the many responsibilities he saw attached to the Director's role:

SHARP: Maybe we could talk . . . [about] just how you saw your role as Director.

REVELLE: Well, I saw it in several different ways. One important way was pushing ocean science on the national and international scene, through the National Academy of Sciences and through the Office of Research and various government committees. . . . Being director of Scripps was a powerful lever for that kind of job.

I was always interested in international cooperation, first with the Japanese and then later elsewhere in the world.

I thought it was also important that I should try to do some science, actual research, and I did that. . . .

Later, in 1957, I got very much interested in the carbon dioxide question. I was very active in the development of the International Geophysical Year. . . . That's where the carbon dioxide program got started. It has now become quite a cottage industry, but at that time it was just David Keeling all by himself.

The other thing I tried to do was to encourage our staff to do their best job of research. I used to walk around the place and talk to people about what they were doing.

I was also concerned, of course, with building the institution, getting more people here and getting more activities, getting more ships, getting more money.[25]

In Revelle's eyes, one of his most important responsibilities was to "push ocean science" through various national and international science advisory committees. These committees proliferated in the 1940s and 1950s (Kevles 1971), and Revelle served on a number of them (see table 5.1). These committee memberships put Revelle in a position in which he could both influence the lines of research the government would support, and frame the program at the Scripps Institution to accommodate the state's needs.

Knowledge Generated: Submarine Warfare (Dependent Variable)

Revelle was an active member of a number of committees that designed oceanographic research as related to underwater acoustics. These included the Department of Defense, Research and Development Board on Oceanography (which he chaired in 1951), and the Joint Chiefs of Staff, Joint Commission on Oceanography (of which he became a member in 1951). In addition, Revelle also was a member of the Department of Interior, Arctic Research Advisory Board (1948–1951). The Arctic was considered a vital strategic area because of its proximity to large areas of the USSR and North America, and because submarines could use the polar ice as cover.

It is not possible to reconstruct precisely what these committees discussed. In two oral history interviews, Revelle (who died in 1991)

Table 5.1
Revelle's Committee Memberships, 1948–1983 (Partial List)

National Research Council (NRC), Pacific Science Board, 1948
NRC, Committee on United Nations Economic and Social Council (UNESCO)
NRC, Committee on Amphibious Operations
Department of Interior, Arctic Research Advisory Board, 1948–1951
National Academy of Sciences (NAS), Committee on Oceanography, 1949–1951
American Geophysical Union, Oceanography Section, vice president, 1950, later president
Department of Defense, Research and Development Board on Oceanography, chair, 1951
Joint Chiefs of Staff, Joint Commission on Oceanography, 1951
U.S. National Committee, International Geophysical Year, Panel on Oceanography, chair, 1957
NAS, elected 1957
U.S. National Commission for UNESCO, 1958
International Council of Scientific Unions, Special Committee on Oceanic Research [later Scientific Committee on Oceanic Research], first president
American Association for the Advancement of Science, president, 1973
NAS, Energy and Climate Panel, 1977, prepared report for Council on Environmental Quality on the "greenhouse effect," 1979
NAS, Panel on Climate Change, issued report, *Changing Climate,* 1983

does not offer many details. But there is considerable evidence that decisions made in these committees shaped marine science at Scripps in the postwar era.

In this era, traditional lines of research expanded, and new ones appeared. For example, research in undersea acoustics, in conjunction with antisubmarine warfare and rescue operations at sea, expanded in the postwar years. Wartime research revealed that at depths between 1,500 and 4,000 feet, undersea currents carried audible sounds for long distances. By the late 1940s and early 1950s the Navy began to develop techniques to transmit signals made by depth charges (small TNT explosions) through these channels to hydrophones (Hackmann 1984, 20). To develop this technique, called Sound Fixing and Ranging (SOFAR), required better knowledge of the movement of sound through the ocean.

By the early 1950s, the United States had begun to deploy an extensive network of undersea hydrophones, which became the backbone of acoustic surveillance. The network was known as the Sound Surveillance System (SOSUS). Part of SOSUS was deployed parallel to the Kamchatka peninsula, and another part between Bear Island and the edge of the Barents Sea in the North Atlantic (Hackmann 1984, 355; Sontag and Drew 1998, 50). To interpret information from the hydrophones, the Navy needed basic data on the movement of sound in the oceans.

By the mid-1950s, advances in submarine technology created a demand for further research in undersea acoustics. Sonar developed during the war had been reasonably accurate at locating diesel-powered subs. Longer-ranging sonar (using low-frequency sound) was required to develop methods to track nuclear-powered submarines.

The Navy invested heavily in long-term research to improve submarine communication and surveillance. Of particular interest were studies of the circulation of water masses of different temperatures; marine organisms such as plankton, fish, and marine mammals; and the topography of the sea floor. Each of these could reflect and refract sound, thereby affecting acoustic signaling and surveillance.

The Navy contracted research in underwater acoustics to a number of laboratories. At Scripps, MPL developed a multifaceted program including work on ocean currents, surface waves, water temperature and density, geology, geophysics, and geochemistry. The laboratory initiated a broad survey of the properties of underwater sound under a variety of oceanographic conditions.

Considerable effort was focused on oceanographic conditions in the Arctic. The Navy regularly deployed submarines off Vladivostok and in the Barents Sea, in an attempt to gather the latest information about Soviet submarines and missiles (Sontag and Drew 1998, 50). A program of oceanographic research in the Arctic was also necessary, according to the Department of Defense, to counter "an intensive, systematic Arctic research program" by the Soviets.[26] Initiated in the early 1950s, the Soviet program was based at the Pacific Oceanological Institute in Vladivostok after 1964.[27] Stimulated by this race to explore the Arctic, the Scripps Institution sent the *Horizon* to survey the sea floor in the Gulf of Alaska in 1951, in an expedition dubbed "Northern Holiday." In the mid-1950s, the Scripps Institution did additional surveying in the Gulf of Alaska (the "Chinook" expedition), and in the Bering Sea (the "Mukluk" expedition).

Knowledge Generated: Expeditions (Dependent Variable)

In a number of advisory positions, including the Department of Defense (DOD), Research and Development Board on Oceanography and the Joint Chiefs of Staff (JCS), Joint Commission on Oceanography, Revelle helped to plan oceanographic support for the atomic tests. At the same time, he oversaw the development of a broad program in undersea acoustics related to submarine warfare.

Revelle's advisory positions with the JCS and DOD also put him in a position to expand military applications of work done on the institution's large-scale expeditions. A number of expeditions were launched in the 1950s, sponsored by the University of California with support from the Navy and the National Science Foundation. The most important of these were the Mid-Pacific Expedition (MidPac) and Capricorn. The expeditions employed two ships, Scripps' *Horizon,* and the Navy'*s PCE (R)- 857*. MidPac included both a geophysical study of the Bikini Islands as a followup to Crossroads and a more general program to map the seabed.

The Capricorn expedition, in the fall of 1952, stopped at Enewetok before heading to the South Pacific. The first part of the expedition included scientific support for Operation Ivy. The expedition featured further studies of underwater sound. Of particular interest were studies of the deep scattering layer.[28] The scattering layer is a mass of plankton that, under certain conditions, reflects sound as if it were a "false sea bottom." Martin Johnson found that the movement of the scattering layer could be predicted based on variation in ocean temperature at different times of day.

New Knowledge: Thermonuclear Tests (Dependent Variable)

As chair of the DOD Research and Development Board on Oceanography, and as chair of the JCS Joint Commission on Oceanography, Revelle would have been in a position to organize the oceanographic research program connected with Operation Ivy. Ivy, scheduled for November 1952, was the first test of a thermonuclear weapon (Gaddis, 1997, 110–111). Revelle and a team of physicists from Scripps, including Munk and Isaacs, participated.

Operation Ivy's purpose was to detonate a device designed by Edward Teller and Stanislaw Ulam at Los Alamos. The weapon's yield was equivalent to ten million tons, or ten megatons, of TNT. It was one thousand times more powerful than the bomb dropped at Hiroshima (York 1976, 82; Rhodes 1995, 511). Physicists called it the "super."

From previous survey work at Bikini, Munk, Isaacs, and Revelle were aware that the atomic explosions had caused part of the coral reefs to slough off. They were concerned that the thermonuclear blast would cause a submarine landslide, which in turn would trigger a tidal wave. A very large wave would flood low-lying areas where people lived. They planned to measure the wave height using a number of methods and provide early warning if necessary to evacuate islanders.

The oceanographers' research plan went awry. A Navy meteorologist, by mistake, had positioned the team dangerously close to ground zero. As Munk later recalled, radioactive spray and debris rained down on them. Munk and his colleague scrambled toward the *Horizon,* which was anchored directly under the fallout (Morgan and Morgan 1996, 49). The vessel remained radioactive for years afterward.

In the decade that followed, Scripps provided research support to a number of other tests. In 1954, Alfred B. Focke of MPL was appointed scientific director of Operation Wigwam. Four vessels from the Scripps Institution, *Spencer Baird, Horizon, Paolina-T,* and *T-441,* supported the operation. The team began with a series of aerial surveys in the spring of 1954, in preparation for the test in May 1955. These were among the first systematic studies of the uptake of fission products and their concentrations into the marine food web (Shor 1978, 405–406).

New Knowledge: Climate Change (Dependent Variable)

Stimulated in part by the atomic and thermonuclear tests in the Pacific, Revelle and his colleagues at Scripps initiated a new program to study weather and climate, funded by the Atomic Energy Commission and ONR. Some of these projects can be seen as the work of Revelle, Seuss, Keeling, and Munk working qua individual scientists, albeit in coordination with one another.

For example, Revelle and Hans Seuss initiated a program to monitor ocean temperature, carbon dioxide concentrations, and oceanographic conditions in the Arctic. They developed the calculations and models to explain oceanographic processes that affect the uptake of CO_2 in the oceans (Malone, Goldberg, and Munk 1998; Hart and Victor 1993, 23).

In addition, Revelle brought Charles Keeling to the Scripps Institution in 1957. Keeling began a series of measurements of CO_2 concentrations in sea water from a monitoring station in Mauna Loa, Hawaii. These measurements, taken continuously since 1957, were a milestone in carbon cycle research. They document continuously increasing CO_2 concentrations from 1957 to the present. Revelle began to express concerns that increasing atmospheric concentrations of CO_2 as a result of industrial emissions was a "global geophysical experiment" (Malone, Goldberg, and Munk 1998, 10).

Climate change, and in particular the role the oceans play in the carbon cycle, continued to concern Revelle and his colleagues in the years that followed. Revelle and Charles Keeling published a paper in 1975, demonstrating that atmospheric carbon dioxide had increased by 5 percent in the previous twenty years. Munk and Revelle published a paper indicating that one third of the increased carbon dioxide in the oceans originated in the clearing of forests for cultivation.

Some of these projects, on the other hand, can be seen as being done by individuals working qua group members. For example, in 1977, Revelle chaired an NAS panel created to assess the possible consequences of increased CO_2 concentrations on the climate. The NAS report, issued in 1983, was titled *Changing Climate.*

The work of Revelle and his colleagues, beginning in the 1950s and 1960s, also stimulated large-scale, collaborative research projects on climate change. These projects include an institutional dimension to the extent that a common framework shapes the research, and different parts of the program are divided among different parts of the team.[29] According to Malone, Goldberg, and Munk (1998), Revelle became involved with one of these projects, the International Geosphere-Biosphere Program (IGBP) through his work with the International Council of Scientific Unions, Scientific Committee on Problems of the Environment. Revelle suggested the original goal for the IGBP (in 1986): "to describe and understand the interaction of the great global physical, chemical, and biological systems regulating planet Earth's favorable environment for life, and the influence of human activity on the environment" (Malone, Goldberg, and Munk 1998, 11).

New Knowledge: Radiation Ecology (Dependent Variable)
The atomic and nuclear tests in the Pacific not only stimulated early work on climate change but also generated a new subfield: radiation ecology.[30] Beginning with the tests in 1946 and the resurvey in 1947, the military brought in marine ecologists to survey the diffusion of radionuclides in the ocean. In 1954, Revelle secured a grant of $1 million from the Rockefeller Foundation for marine biology, much of it radiation ecology.[31] Revelle worked with a number of policy groups to translate research into recommendations for curtailing release of radioactive material into the oceans. The Scripps Institution advised the National Commission on Radiological Protection, the International Council on Radiological Protection, and the UN International Atomic Energy Agency. Revelle later chaired the Oceanography and Fisheries Panel of the National Academy of Sciences Committee on Biological Effects of Atomic Radiation (BEAR).

This was one of a number of NAS committees studying the effects of atomic energy. The committee that Revelle chaired met twice in the spring of 1956. Its summary report was published shortly thereafter. Detlef Bronk, President of the NAS, summarized the process as follows (NAS/NRC 1957):

> Rough drafts of most of the materials published [in the summary] were prepared at the second meeting. These reports, which give the detailed technical background of the Committee's findings and recommendations, have been completed during the past years. . . . All the members of the Committee participated in planning and outlining the materials covered. . . . After the publication of its Summary Report in June 1956, the Committee . . . met informally with scientists from the U.K. [in September 1956]. The discussions centered around the recommendations that could be made on the basis of existing knowledge and the nature of research needed in planning the disposal of radioactive wastes at sea . . . the facts upon which the study's conclusions are based result from more than two decades of research.

As Bronk and others reveal, the committee was formally constituted by the NAS, and Revelle was appointed chair. The group met twice. At each meeting, members presented their research on various aspects of the problem at hand, that is, the effects of radiation on the marine environment. Each member of the group participated in planning and outlining the report. Revelle (together with Milner B. Schaefer, Director of the IATTC) wrote an introduction to the report (NAS/NRC 1957), which summarized the main findings. The committee warned,

Waste products from nuclear reactions require special care: they constitute hazards in extremely low concentrations and their deleterious properties cannot be eliminated by any chemical transformations; they can be dispersed or isolated, but they cannot be destroyed. Once they are created, we must live with them until they become inactive by natural decay, which for some isotopes requires a very long time.

Revelle's position with the Navy and his role with JTFO situated him as one of the first scientists to design a systematic study of the effects of radiation on ecosystems. His memberships on a number of scientific planning committees, most notably with the NAS, the National Committee for the International Geophysical Year, and the International Council of Scientific Unions, positioned him as one of the more prominent members of this new subfield. Less than a decade later, Revelle also had a hand in organizing one of the most important international, ecosystemic research programs, the International Biological Program (IBP). Revelle chaired the U.S. national committee of IBP in 1964.[32]

Radiation ecology provides a second illustration of how hybrid institutions generated new areas of research. They set in motion research activity that led in directions that an earlier generation of scientists could not have expected. The U.S. role generated the atomic tests, which generated concerns about radiation in marine ecosystems, which stimulated more money for marine ecology. Ritter looked at the Pacific and saw a set of questions about marine life that, set end to end, were potentially infinite. In 1915, had he been able to compare the work of his biological station with this new field of "radiation ecology," he would most likely have found this new subfield, and the questions it raised, to be somewhat alien. Radiation ecologists in the 1950s and 1960s were primarily concerned with the effects of radiation on the marine food chain. Like their predecessors at Scripps, they continued to be concerned with marine life, together with the chemical and physical properties of the ocean. Unlike Ritter, their questions were much more constrained by concerns about the circulation of radionuclides.

Fisheries Research (Dependent Variable)

Of course, the U.S. secondary role did not redirect *all* the Scripps Institution's research activity. Some new lines of research can be better explained more conventionally, with reference to individual self-interest and collective action. The Marine Life Research (MLR) program, part of the State of California's California Cooperative Oceanic Fisheries Investigations (CalCOFI), is an example.

The pressures that motivated the State of California to organize MLR, beginning in 1947, are reminiscent of those bearing on the U.S. Fish Commission when it funded Spencer Baird's laboratory at Woods Hole, or those that motivated European governments to fund ICES. In California, the sardines had disappeared. In 1936-37, fishermen landed 726,000 tons; in 1946-47, 234,000 tons; and in 1947-48, 130,000 tons. People who depended on the sardine for a living, including some 3,000 fishermen and people connected with the hundred canneries in Monterey Bay, Los Angeles, Newport, and San Diego, wanted to know why the fish had disappeared. Was it overfishing? altered migration patterns? the atomic bombs?

In response to this sardine crisis, the California legislature created a cooperative research program. CalCOFI brought together researchers from the U.S. Fish and Wildlife Service, the California Division of Fish and Game, the California Academy of Sciences, Stanford University's Hopkins Marine Laboratory, and Scripps. The legislature allocated $300,000 over three years (a sum that was later increased) to MLR.

MLR can be seen as a continuation of work Sverdrup had begun before the war on the California Current. Researchers were interested in correlating changes in oceanographic conditions with the maximum areas of sardine spawning. After ten years, the program had accumulated considerable data on the California Current, including coordinated physical, chemical, and biological observations. They identified areas of spawning and migration patterns. As a result, the California Current is one of the best-documented ecological systems in the world, and the sardine one of its best-known inhabitants.

Still, if anyone had hoped MLR would bring back sardines in the numbers reminiscent of the 1930s, they would have been disappointed. Researchers examined layers of sediments and counted the remains of fish scales of different types. Numbers derived from scales preserved in the sediments were used to estimate the abundance of various species each year. Researchers estimated that the sardine numbers of the 1930s represented an 800-year high.

After the war, the United States funded a number of other new scientific research programs directed toward expanding Pacific fisheries. The

Pacific Ocean Fisheries Investigation (POFI), based in Hawaii, was the first. As part of this effort, Revelle took the lead in persuading U.C. President Sproul to establish an intercampus institute for the study of commercial fisheries. The Institute for Marine Resources (IMR) was Revelle's effort to bring the Scripps Institution into the network of agencies then developing for the study of Pacific fisheries (Day 1985, 17).

5.3 Alternative Explanations

The postwar U.S. role of hegemon profoundly reshaped the knowledge the Scripps Institution generated. During the Cold War era, the Director of the Institution served on a number of advisory committees of the NRC, DOD, and JCS. Through these positions, the new hegemon's purposes and goals transformed the knowledge the Scripps Institution generated. The mechanism was the positional fix.

By contrast, the neorealist model would predict that the United States would fund military applications of oceanography consistent with its national security interests. Again, the model derives interests from a state's position in a hierarchy based on military capabilities (Waltz 1979). As a great power (or hegemon) the United States would seek to increase its power (or security) by funding defense-related oceanography during the war and throughout the Cold War era.

In previous chapters, a neorealist explanation could be sharply differentiated from an institutional one and rejected. In earlier decades, the United States possessed sufficient military capabilities to compete as a great power. To neorealists, this means that the United States would fund military applications of oceanography at a constant rate, from about 1890 to 1941.

In periods in which the United States had capabilities to exercise a role as great power (or hegemon), but rejected the role, the institutionalist prediction differs from the neorealist one. Such periods include 1905–1917 and the 1920s and 1930s. Evidence supports the institutionalist interpretation, not the neorealist one. When the United States rejected a role as great power (embraced isolationism), the Scripps Institution did not conduct research on military applications of oceanography *despite preponderant U.S. capabilities*. Research practices changed as a result of changes in the U.S. role (mediated through channels like the

NRC), *not* as a result of preponderant U.S. capabilities (which can be seen as roughly constant from 1890 to 1941).

For the period from 1941 to 1970, the neorealist explanation cannot be as clearly differentiated from an institutional one. The United States had preponderant capabilities to act as a great power, *and* it accepted the role. In this period, neorealism and institutionalism both predict a shift away from basic oceanography (like Sverdrup's survey of the California Current) toward military applications (like thermonuclear tests).

At the same time, the chapter provides additional observations on the variables of concern (the U.S. role, the Director's role, and knowledge generated). The specific rights and responsibilities attached to the role of great power (during the war) differed from those attached to the role of hegemon (after the war). Framed and reframed by these changing responsibilities, the research program at Scripps changed during and after the war. Such evidence suggests that an institutional explanation is both plausible and more nuanced than a neorealist one and is worthy of further research.

In the post–World War II era, there is some limited evidence to support the interest group approach. Although the Scripps Institution's main program was not penetrated by business interests, MLR was created primarily to meet the needs of the fishing industry. In other words, interest group lobbying can explain the creation of MLR, a research program intended to benefit California's commercial fisheries.

5.4 Summary

This chapter provided two sets of observations on the explanatory and dependent variables. First, the United States assumed a role as great power during the Second World War, supporting its allies in a global struggle (explanatory variable). Revelle assumed a role within the Navy's Bureau of Ships. Sverdrup, as Director of Scripps, was largely denied an active role in military research because he was denied security clearance. Research results included work on ocean waves and surf, the effects of oceanographic factors on amphibious landings, and submarine warfare (dependent variable).

Second, the United States accepted a role as global hegemon (competing with the USSR) in the early Cold War era (explanatory variable).

Mediating between the new U.S. responsibilities and the Director's role were an expanding web of military advisory committees. The rights and responsibilities of U.S. hegemony reshaped postwar research. Submarine warfare was expanded, and new lines of work, including thermonuclear tests, carbon cycle research, and radiation ecology, were added (dependent variable).

Crucially (and unpredictably) the U.S. role as hegemon was generative of new lines of research, including new research on climate change. Had the Director (Revelle) not been involved in planning atomic and thermonuclear tests during and after the war, it is unlikely that he would have thought as comprehensively about the effects of human activity on global ecosystems. Most likely, he would have continued with surveys of the more limited ecosystems of the southern California region, consistent with Sverdrup's research agenda in the late 1930s. The U.S. role (mediated through networks like the NRC) reframed oceanographic research toward the study of *global* systems, albeit in directions that cannot be fully determined by political purposes alone.

Shifting away from global politics, the next chapter homes in on fisheries research on Scripps' campus, at the Inter-American Tropical Tuna Commission. Unlike chapters 3–5, which documented the effects of global political changes on Scripps' research program, chapter 6 focuses on the effects of international treaty rules on fisheries research (particularly tuna and dolphin abundance). To further illustrate how institutions shape the generation of new knowledge, it emphasizes two mechanisms: the statutory fix and the committee fix.

6

Inter-American Tropical Tuna Commission, 1950s–1990s

We need to get the information cooperatively with the Latinos, so they will believe us.

—W. B. Chapman, U.S. State Department, in a letter to Montgomery Phister, American Tunaboat Association, 1949[1]

In October 1950, Milner B. Schaefer, who expected to be appointed as first Director of the Inter-American Tropical Tuna Commission (IATTC), wrote to Roger Revelle, asking him to find space for the IATTC on the Scripps campus. Revelle agreed that such an arrangement would be mutually beneficial to the Scripps Institution and to the IATTC.[2] Although the IATTC was an international organization, with no formal connection to Scripps, the marine scientists at the two organizations developed a working partnership. Previous chapters focused on changes in the Director's role over time, and its impact on the research program. This chapter focuses on two additional mechanisms: the statutory fix and the committee fix. The statutory fix refers to ideas or epistemic frameworks, specified in formal or informal rules, that guide research activity. The committee fix refers to standardized procedures, like annual meetings, through which an organization regularly accepts beliefs on matters of fact.

In the period under consideration (1950–1990), the United States continued to exercise hegemony, particularly in the geographical area of interest, the Western Hemisphere. For purposes of this case study, the power structure (as neorealists define it) can be considered a constant. Crucially, power structure cannot explain the observed changes in the dependent variables. It is possible, therefore, to more clearly compare an institutional approach with an alternative explanation based on interest group activity.

This chapter excludes elements that appear in other, more general, histories of the "tuna-dolphin conflict." For example, it does not review the history of economic sanctions the United States imposed on exports of tuna from a number of Latin American states. Nor does it explain how nongovernmental organizations worked to put dolphin protection on the international agenda. Interested readers should consult other sources (see, e.g., NRC 1992; Parker 1999). Setting these elements aside permits a more detailed analysis of mechanisms through which institutions shape the generation of new knowledge and thereby also regulatory action. To develop the institutional argument as clearly as possible, and to compare it to an alternative based on the activity of interest groups, the chapter's time frame extends to the 1990s.

Sections 6.1 and 6.2 clarify the argument. Section 6.3 develops a structured, focused case history of the IATTC's research and regulatory program (King, Keohane, and Verba 1994). It presents a history of research and regulatory action on tuna fishery and on the industry's impact on dolphins. Sections 6.4 and 6.5 explain the limits of an explanation framed purely in terms of interest groups, and the value added by an institutional approach.

6.1 How Do Organizations Form Beliefs?

This chapter makes a causal argument, divided into two parts. The first part concerns how organizations translate changes in phenomena (e.g., fish or dolphin populations) into changes in knowledge. It is assumed that phenomena existed and actually changed, regardless of the IATTC's beliefs about them. The question is how the IATTC generated, selected, and used certain statements about these phenomena to guide regulatory action.

The argument is that agents embed particular ideas in institutional rules. Guided by these frameworks, the organization's scientific staff generates new knowledge or statements (statutory fix). At the organization's annual meetings, negotiators accept some of the scientific staff's findings temporarily, for purposes of discussion. Also at annual meetings, the Commission formally accepts some of the scientific staff's findings to ground or to justify action (committee fix). The institutional argument is that, other things equal, *organizations use the statutory fix and the committee fix as institutional selection mechanisms to translate changes in phenomena into changes in knowledge.*

Another possibility is that change in the relative strength of interest groups better explains the observed change in the IATTC's knowledge. This approach specifies the fundamental interests of societal actors exogenously and assumes they are fixed for purposes of analysis (Moravcsik 1997). In general, individuals and firms seek to maximize net income in a given period (Milner 1997). Nongovernmental organizations (NGOs) maximize attainment of their goals, for instance, to protect fish and dolphins. The interest group argument is that, other things equal, *when the relative strength of interest groups changes, the organization's knowledge tends to change.*

6.2 From Knowledge to Action

The second part of the causal chain concerns the mechanisms by which new knowledge is brought to bear on policy action. Here the dependent variable is regulatory action under IATTC auspices from about 1950 until the present.

Again, two possible explanations are considered. First, changes in the IATTC's knowledge over time drove changes in its recommended regulations. More specifically, *as the IATTC's beliefs about the degree of anthropogenic threat to certain species increased, the extent of regulatory protection changed (from weaker to stronger).* Second, change in the relative strength of interest groups over time caused change in the IATTC's regulatory action. *As the relative strength of environmental NGOs increased, regulatory protection changed (from weaker to stronger).* Alternatively: *As the relative strength of firms declined (compared to environmental NGOs), regulatory protection changed (from weaker to stronger).*

Although the argument is presented as two parts of a causal chain, the case history is not presented this way. To preserve the coherence of the narrative, section 6.3 proceeds in roughly chronological order. It reviews the history of research and regulation for yellowfin, from about 1950 to the present. It then summarizes the history of research and regulation for dolphins, from the late 1970s to the present. At the same time, the narrative focuses on the explanatory and dependent variables. Analysis of the arguments is deferred until sections 6.4 and 6.5.

6.3 Case History: IATTC's Scientific Program, 1950–Present

Conservation Embedded in Treaty Rules

The Convention that created the IATTC was one of the earliest international treaties that explicitly embedded conservation as a management goal.[3] Since 1950, more than thirty international commissions have been created to deal with actual and potential conflicts over highly migratory, anadromous, and straddling stocks. Of these, a number produce scientific research, with conservation as an organizing framework (FAO 1996). These include the International Commission for the Conservation of Atlantic Tunas, the South Pacific Commission, and the Indian Ocean Tuna Commission. The regime of Exclusive Economic Zones (EEZ), as formalized in the 1982 UN Convention on the Law of the Sea (UNCLOS), also embeds conservation (UNCLOS Article 61).

The Convention that created the IATTC embeds conservation in Article II. It states (in part):

> The Commission shall perform the following functions and duties:
>
> Make investigations concerning the abundance, biology, biometry, and ecology of yellowfin (*Neothunnus*) and skipjack (*Katsuwonus*) tuna in the waters of the eastern Pacific Ocean fished by the nationals of the High Contracting Parties . . .
>
> Collect and analyze information relating to current and past conditions and trends of the populations of fishes covered by this Convention.
>
> Study and appraise information concerning methods and procedures for maintaining and increasing the populations of fishes covered by this Convention . . .
>
> Recommend from time to time, on the basis of scientific investigations, proposals for joint action by the High Contracting Parties designed to keep the populations of fishes covered by this Convention at those levels of abundance which will permit the *maximum sustainable catch*.[4] [emphasis added]

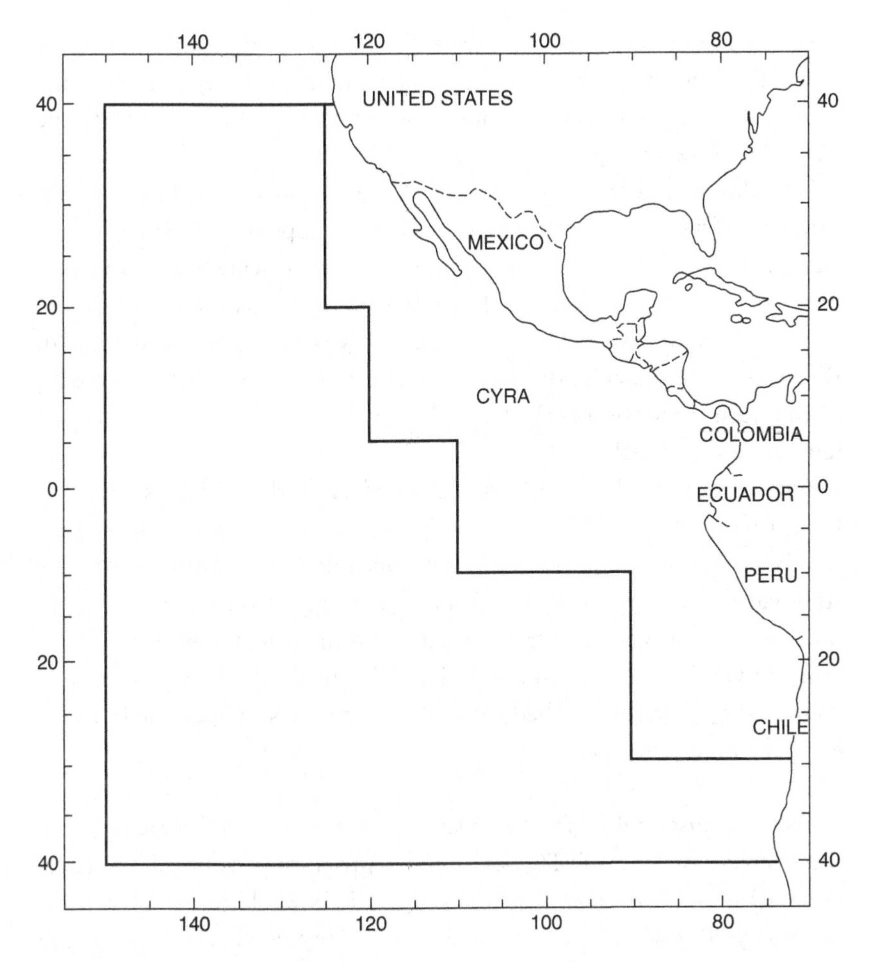

Figure 6.1
Map of Convention Regulatory Area. *Outer area*: dolphin regulation, per 1998; *inner area:* yellowfin regulation, per 1950 agreement (CYRA).
Source: IATTC, Annual Report (1998), 93.

The Convention was designed to create a commission with a scientific staff that would produce knowledge to assist the industry in its effort to maximize sustainable catches of tuna and baitfish (committee fix). Article II (5) of the treaty assigned the IATTC the responsibility to research the yellowfin (*Neothunnus*) and skipjack (*Katsuwonus*) in the eastern Pacific Ocean, and to recommend regulations to member states, if necessary.[5] By the late 1950s, the IATTC's scientific staff became an influential epistemic community, and its findings shaped international regulation in this area of the Pacific Ocean. The area of concern to the IATTC is shown in figure 6.1.

Why did the IATTC's scientific staff prioritize models emphasizing conservation of fish rather than a more ecosystemic approach to management? At that time, marine scientists, including Milner B. Schaefer, the IATTC's first Director of Investigations, knew that marine ecosystems were complex, and factors other than fishing effort would affect abundance. In February 1951, Schaefer outlined a program of research and presented it at an Executive Session of the IATTC.[6] It included a rather broad, ecosystems-based approach.

Several months later, Schaefer learned that the IATTC's research budget would be less than one third of the amount he considered necessary (Bayliff 2001, 55). Therefore, Schaefer told the Commission that his staff was not able to proceed as planned. He suggested that his team confine itself to that work the treaty required, estimating the maximum sustainable yield (MSY) for tuna and baitfish.[7] In short, the framework of conservation, formally embedded in the treaty rules, shaped the IATTC's early research program.

Maximum Sustainable Yield Models Consistent with Article II (5) of the Convention, the IATTC developed a program of research to determine maximum sustained catches. The IATTC's work on maximum sustainable yield was very influential among fishery managers beginning in the 1950s. Schaefer developed a model that could predict MSY using catch and effort data supplied by the fishermen. He was influenced by earlier work by Thompson and Bell (1934), who used data derived from the Pacific halibut fishery in the 1920s. Thompson and Bell found that fishing was inversely related to catch per unit of fishing effort. Schaefer was also familiar with a paper by Hjort (1933) on whale populations that suggested that population productivity was related to population size. Schaefer's contribution to fisheries biology was to develop a model for MSY using data that could be obtained fairly easily from fishermen's logbooks and canneries (Smith 1994, 255).

The Convention directed scientists to investigate, among other things, the abundance of yellowfin. Schaefer's model used catch per day's fishing

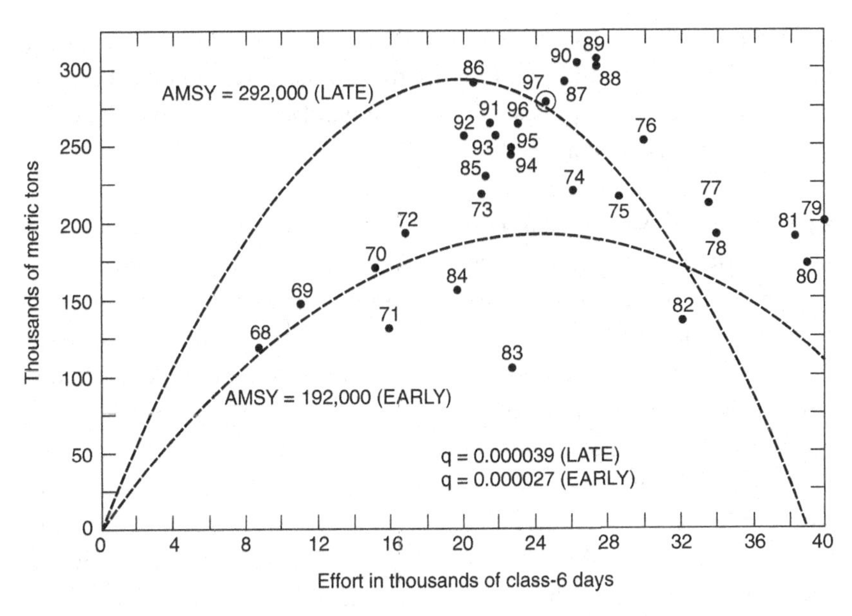

Figure 6.2
Estimates of Maximum Sustainable Yields, 1967–1997. Each point represents an estimated relationship between effort and yield.
Note: AMSY = average maximum sustainable yield.
Source: IATTC, Annual Report (1997), 116, fig. 45.

(CPDF) to indicate yellowfin abundance. He assumed a symmetrical yield curve for yellowfin, an inverted U shape. With zero fishing effort, yield was zero, increasing smoothly to a maximum, declining thereafter to some extreme level of effort that theoretically would yield zero yellowfin. The mathematics of Schaefer's model and its impact on the science of fisheries biology are detailed in Clark (1976, 15–16) and NRC (1997, 38).

Figure 6.2 shows data for catch and effort from 1967 to 1997. Average maximum sustainable yield (AMSY) was estimated to be between 181,000 metric tons and 298,000 metric tons of yellowfin. In the late 1970s and early 1980s the fishery was overcapitalized (Gordon 1954). Both the surface catch and CPDF had declined compared to the mid-1970s. In 1982-83 (an El Niño year), apparent yellowfin abundance, catch, and effort declined. Purse seiners took in less than the recommended quota of yellowfin. Thereafter, many vessels shifted to the Atlantic or western Pacific. From 1985 to 1997, total catches remained roughly within the area defined as AMSY. Each of the points in figure 6.2 represents a statement, generated by the scientific staff, about the estimated relationship between effort and yield. When published in the Annual Report, the point (and the statement) can be seen as the IATTC's group belief.[8]

Age-Structured Models and Yield-Per-Recruit Based on the embedded framework of conservation, the IATTC also developed age-structured and yield-per-recruit models. The models describe a cohort of fish over its lifetime. A cohort refers to all the fish recruited into a fishery at the same time, for example, in a six-month period. Estimates are made of the numbers of fish in each cohort at the time of recruitment (the entrance of young fish into the fishery) and at various intervals thereafter. Estimates are taken of the number and average weights of fish in each cohort in the catch, an estimate of natural mortality rate, and an estimate of the mortality rate due to fishing. These estimates, particularly of mortality, are somewhat crude.[9] From this set of measurements, indices of abundance are constructed for the stock as a whole and for larger fish (e.g., those that have been in the fishery for more than seventeen months).[10] The basic mathematics of age-structured models, developed by Beverton and Holt (1957), are explained in Smith (1994, 315–323) and Clark (1976, 269–292).

Yield-per-recruit (YPR) models use cohort data to estimate abundance. Yield-per-recruit refers to the total yield in weight harvested from a cohort of fish over its lifetime, divided by the number of fish in that cohort. When a cohort of fish is young, its rate of growth is highest because each fish grows rapidly. As it matures, the cohort grows more slowly, and it eventually disappears. The management goal derived from yield-per-recruit analysis is to harvest the fish, on average, at about the critical size. For a stock of fish, the critical size is determined by identifying the point at which the loss in total weight to the stock by natural mortality balances the gain to it by growth. Fishery scientists believe that if effort is directed to fish of critical size, on average for the fishery as a whole, the yield tends to be larger (other things being equal). Diminished yield-per-recruit indicates growth overfishing, usually due to overcapitalization.

In general, the management goal is for about half of the fish caught to be less than the critical size and about half larger than the critical size. Growth overfishing, or taking a larger proportion of fish that are too small, indicates overcapitalization. This became a problem in the eastern Pacific by the late 1960s and 1970s. In general, to limit growth overfishing, fisheries managers set guidelines for minimum size limits for fish, acceptable mesh sizes for the fishing nets, or restrictions on fishing times or areas to prevent capture of undersize fish.[11]

Figure 6.3 illustrates the estimated yield using yield-per-recruit and production models. The dashed lines show the estimated maximum yield of yellowfin in weight (derived from the yield-per-recruit model). The solid lines show the maximum yield using production models (described in the next section). Actual yields for 1968–1983 (upper panel) indicate growth overfishing, which was alleviated after 1982-83, an El Niño year. Since 1984, the fishery has continued to operate at lower levels of capitalization as compared to the late 1970s.

Production Models Using Catch and Effort Data As shown in figure 6.3, the IATTC has also developed production models to estimate yellowfin yield. Production models use estimates for the fish population as a whole. Biologists assume that catches in the eastern Pacific Ocean (east of 150 degrees west longitude) are from a single stock.[12] Catch divided by effort is used as a rough estimate of the population. This is based on an assumption that CPUE (or CPDF) is proportional to the stock level (Clark 1976, 14). The IATTC also uses a second type of production model in which the relation between catch and effort need not be proportional. The parameters for production models are estimated with regression techniques, using catch and effort data for a number of years.[13]

In general, rates of recruitment, growth, and natural mortality are sensitive to predation and to environmental factors. When fishing is introduced into the population, these rates change. Rates of recruitment and growth increase, and population gains exceed natural mortality by a certain margin. Biologists assume, when fishing is introduced, catch per unit of effort is proportional to the stock size. Changes in the rates of recruitment and growth due to small increases in effort would more than offset the decreases in catch per unit of effort (and stock size). At higher levels of effort, catch per unit of effort and stock size would decline with increased effort. The goal of using production models is to identify

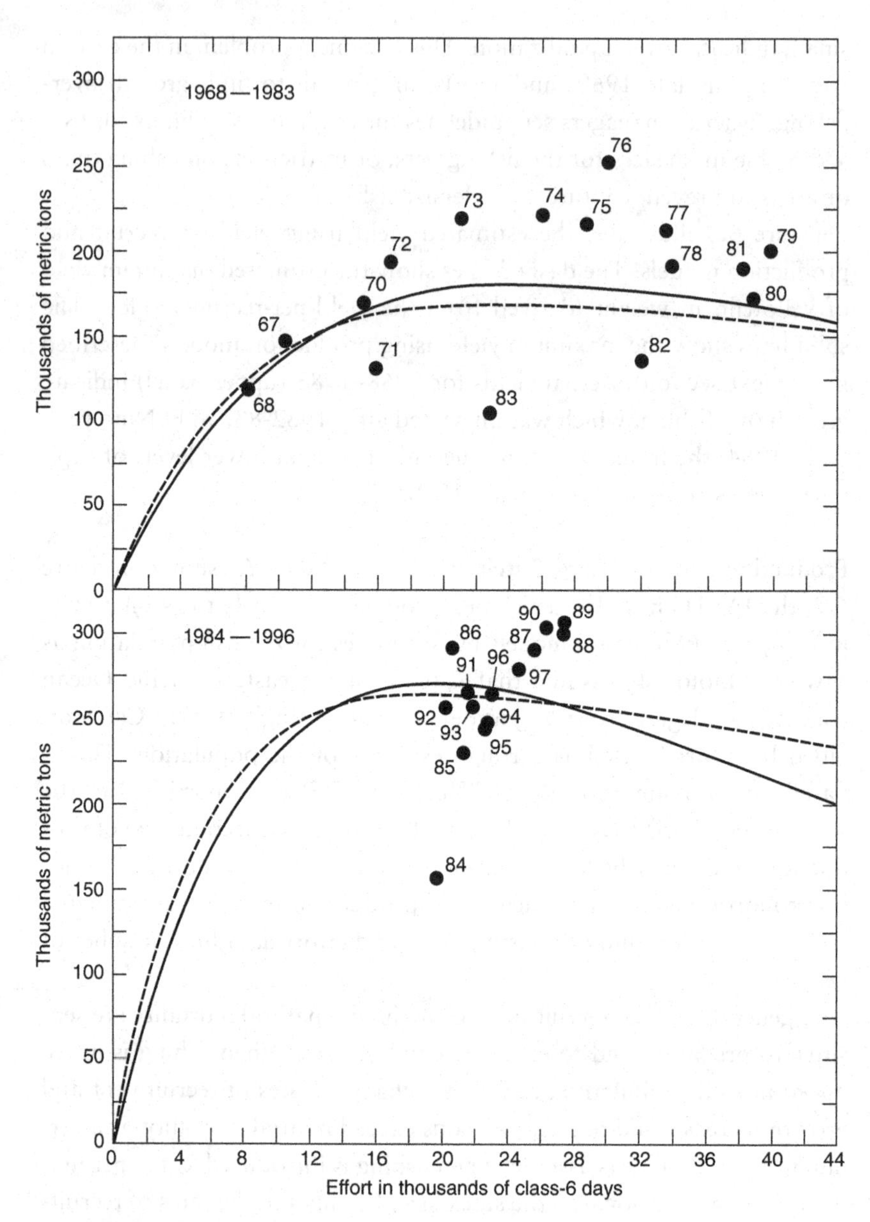

Figure 6.3
Tuna Abundance, Yield-Per-Recruit, and Production Models, 1968–1983 and 1984–1996. Yields of yellowfin the EPO corresponding to conditions during the 1968–1983 and 1984–1997 periods estimated from YPR (dashed curves) and production models (solid curves).
Source: IATTC, Annual Report (1997), 120, fig. 50.

an optimum size for the stock as a whole that will maximize yield from the fishery.[14] Each data point in figure 6.3 represents a statement about the estimated relationship between effort and yield. When published in the IATTC's Annual Report, these statements can be understood to represent group beliefs.

From Knowledge to Action: Fishery Regulations

Unlike many international fisheries, in the 1940s the yellowfin fishery in the eastern Pacific did not resemble a tragedy of the commons (Hardin 1968) or a Prisoner's Dilemma failure of collective action (Axelrod and Keohane 1986). Rather, yellowfin were believed to be highly abundant, and coastal states were eager to lay claim to what they perceived to be the ocean's rich bounty.

The belief that tuna were abundant, if accepted by the major fishing states in the region, would legitimize further fishing by U.S. flag vessels in areas off the Pacific coast of Mexico, as well as other Central and South American states (Scheiber 1988, 464–469). In the early years, the scientific staff generated data that demonstrated fishing effort for yellowfin was low enough, and yields high enough, to support further increases in capital.

At the IATTC's meeting in 1957, for example, the Director reported that the staff found skipjack abundance to be "well below MSY" and that yellowfin abundance was "not far below MSY."[15] That year, the IATTC decided not to regulate. It accepted the knowledge the staff produced, by means of a formal resolution. The Commissioners accepted the following language:

> The intensity of fishing on the tuna and bait fish stocks . . . *remains at a level sufficiently low that none of the species with which we are concerned appear to be in danger of overfishing. Yellowfin tuna are being fished at an intensity not far below that which is estimated to correspond to maximum sustainable average catch,* but with the present capability of fishing fleets, it is very unlikely that such a level of fishing effort can be attained, even with full operation of all vessels. *There is, therefore, no need for conservation recommendations at this time.* [emphasis added]

The IATTC's belief that yellowfin and skipjack stocks were below the point of MSY justified its decision not to regulate. The Commission thereby sent a signal (a "green light") to the industry that it could increase effort.

It was not long, however, before the data began to reflect overfishing. The Director advised the Commission, at the IATTC's February 1961 meeting in Panama City, that fishing for yellowfin was already at or beyond the point of MSY.[16]

The Commission accepted the Director's report. At a special meeting in Long Beach in September 1961, the Commission recommended regulatory action. To illustrate how the IATTC formally accepted this belief and used it to ground its recommendation for action, it is necessary to quote the resolution directly:

> *Observing*, that the studies of its scientific staff have indicated that during the year 1960 the intensity of fishing for yellowfin in the Eastern Pacific Ocean had reached the level corresponding to maximum average sustainable catch;
>
> *Observing*, that continuing studies of catch statistics and other data indicate that, during 1961, there has been a further increase in the amount of fishing for yellowfin tuna, that the total catch during this year will exceed the sustainable yield, and that, consequently, the populations of this species will most probably be reduced to a level which cannot provide sustained maximum yield;
>
> *Concluding*, therefore, that there is a need for joint action by the High Contracting Parties to restore the yellowfin populations to those levels of abundance which will make possible the maximum sustainable yield, and to maintain them in that condition . . .
>
> Having considered that limitation of total catch, by annual quota, is the most effective and practicable type of regulation . . .
>
> *Recommends* to the High Contracting Parties, that they take joint action as follows: Establishment of a quota of total catch of yellowfin tuna by fishermen of all nations . . . during calendar year 1962.

In general, from the early 1960s until the late 1970s, the Commissioners continued to accept scientific statements to the effect that fishing for yellowfin was at or beyond MSY. They also accepted the need for a global quota, a limit on the total number of fish caught (see table 6.1). The Commission flashed a red light. Between 1966 and 1979 (and again in the late 1990s), member states implemented regulations the IATTC recommended (Bayliff 2001, 35–37).

Embedded Preservation and Dolphin Protection

The framework of preservation was embedded in the IATTC's regular program activity in 1976. Environmental activists were determined to reduce or to eliminate dolphin mortality caused by the tuna industry in the eastern Pacific. The United States initiated negotiations, under IATTC auspices, to develop a multilateral program to limit dolphin mortality.[17] The embedding of preservation in the IATTC's rules can be

Table 6.1
Quotas, Catches, and CPDFs (Class-6 Purse Seiners) for Yellowfin, Eastern Pacific Ocean, 1967–1996

Year	Quota	Surface Catch	CPDF
1967	76.7	80.0	6.2
1968	84.4	102.0	14.8
1969	108.9	128.9	15.5
1970	108.9	155.6	12.7
1971	127.0 + (2 × 9.1)	122.8	9.3
1972	108.9 + (2 × 9.1)	177.1	13.2
1973	117.9 + (3 × 9.1)	205.3	11.5
1974	158.8 + (2 × 9.1)	210.4	9.2
1975	158.8 + (2 × 9.1)	202.1	8.3
1976	158.8 + (2 × 9.1)	236.3	9.1
1977	158.8 + (18.1 + 13.6)	198.8	7.3
1978	158.8 + (18.1 + 13.6)	180.5	6.2
1979	158.8 + (18.1 + 1.6)	189.7	5.5
1980	149.7 + (40.8)	158.7	4.9
1981	149.7 + (3 × 13.6)	181.8	5.5
1982	145.1 + (2 × 13.6)	125.1	4.7
1983	154.2 + (2 × 13.6)	94.2	5.1
1984	147.0 + (2 × 13.6)	145.1	8.8
1985	157.9 + (18.1 + 9.1)	217.0	11.9
1986	158.8 + (2 × 13.6)	268.3	15.7
1987	None	272.2	12.6
1988	172.4 + (2 × 27.2)	288.0	12.2
1989	199.6 + (2 × 27.2)	289.4	12.4
1990	181.4 + (5 × 18.1)	273.3	12.6
1991	190.5 + (4 × 18.1)	239.0	13.7
1992	190.5 + (4 × 18.1)	239.8	14.2
1993	226.8 + (4 × 22.7)	232.1	13.1
1994	226.8 + (4 × 22.7)	218.4	12.1
1995	213.2 + (3 × 18.1)	223.6	11.9
1996	213.2 + (3 × 18.1)	249.3	13.0

Source: IATTC, Annual Report (1996), 181, Table 12.
Note: Quotas and catches in thousands of metric tons; CPDFs in metric tons per day. Quotas apply within the IATTC's yellowfin regulatory area. Numbers in parentheses indicate additional quota increment added subject to the Director's discretion.

seen as internationalization of domestic U.S. regulations (DeSombre 2000).

In October 1976, the IATTC agreed to create a program to reduce dolphin mortality. Funding for the program became available in 1978, and the tuna-dolphin program began the following year.[18] Preservation was embedded as follows:

> [The Commission decided at its October 1976 meeting] that the IATTC should concern itself with the problems arising from the tuna-porpoise relationship in the eastern Pacific Ocean. As its objectives, it was agreed that (1) the Commission should strive to maintain a high level of tuna production, (2) also to maintain porpoise stocks at or above levels that assure their survival in perpetuity, (3) with every reasonable effort being made to avoid needless or careless killing of porpoise.[19]

As a result, the IATTC agreed on a new program of study:

> (1) an observer program, (2) aerial surveys and porpoise tagging, (3) analyses of indices of abundance of porpoises and computer simulation studies, and (4) gear and behavioral research and education.[20]

In the 1970s and early 1980s, data on the abundance of dolphins in the eastern Pacific were scarce, and little was known about the tuna industry's alleged adverse impact. In order to justify a decision to regulate (or not to regulate) purse seiners that fished on dolphins, it was necessary to fill this gap in knowledge. To do so, the IATTC began to train observers and to place them on board tuna vessels. The IATTC placed observers on about 15 percent of tuna vessels in the eastern Pacific. Observer coverage was later increased to 100 percent. Data from observers were used to estimate the number of dolphins killed incidentally and to estimate total population sizes (see tables 6.2 and 6.3 and figure 6.4).[21] Here we have a clear example of how a statutory fix led to the generation of a new kind of knowledge.

The data in tables 6.2 and 6.3 and figure 6.4 can be seen as statements the scientific staff made about dolphin mortality. When published in the IATTC's Annual Reports and other documents, the statements can be seen to represent the group's beliefs.[22]

From Knowledge to Action: Regulations to Protect Dolphins

This section focuses on the mechanisms through which new knowledge, generated by the IATTC, contributed to increasingly strict regulation of dolphin bycatch in the eastern Pacific. According to IATTC estimates,

Table 6.2
Annual Estimates of Incidental Dolphin Mortality, Eastern Pacific Ocean, 1979–1998 (Offshore Spotted, Spinner, and Common Dolphins)

Year	No. of Dolphins Killed
1979	21,331
1980	29,752
1981	32,997
1982	28,296
1983	13,448
1984	38,584
1985	58,816
1986	132,169
1987	98,882
1988	81,129
1989	98,451
1990	53,847
1991	27,127
1992	15,539
1993	3,601
1994	4,096
1995	3,274
1996	2,547
1997	3,005
1998	1,877

Source: IATTC Annual Report (1998), 204.

approximately 130,000 animals were killed in 1986, and 100,000 in 1987 (Parker 1999, 12). This rate of dolphin bycatch was reduced by regulation under IATTC auspices, to fewer than 2,000 animals in 1999 (tables 6.2 and 6.3 and figure 6.4).

A first step to reducing bycatch was for the IATTC's scientific staff to generate data documenting the problem and to persuade fishers, NGOs, and states to accept them. A crucial part of this process was the IATTC's effort to place observers aboard purse seine vessels. The IATTC staff, particularly the Director of the Tuna-Dolphin Program, Martin Hall, urged fishers to carry observers, and lobbied their governments to require seiners to carry them. In addition, the U.S. Marine Mammal Protection Act Amendments of 1984 required all vessels in the region to participate in a "comparable" program, which required on-board observation of

Table 6.3
Estimated Numbers of Dolphins Killed Incidentally by the Purse Seine Fishery for Tuna in the Eastern Pacific Ocean, and Percent of Estimated Total Abundance, 1994–1999

Year	No. of Dolphins Killed	Percent of Total Abundance[a]
1994	4,095	0.043
1995	3,274	0.034
1996	2,547	0.027
1997	3,004	0.031
1998	1,877	0.020
1999	1,348	0.014

Source: Bayliff (2001, table 3, 57).
a. The estimate of total abundance of dolphins in the eastern Pacific Ocean, based on pooled data for research vessel surveys conducted during 1986–1990, was 9,576,000 animals.

dolphin mortality. U.S. trade leverage was used to enforce this requirement (Parker 1999, 24; DeSombre 2000). Data collected by observers were used to generate estimates of incidental mortality and stock abundance, and were formally accepted at the IATTC's annual meetings. The IATTC's finding that dolphin bycatch was very high in the 1980s "provided a wake-up call" to the international fleet, which previously had been "in denial" about levels of dolphin mortality.[23]

Another element was a series of workshops in which the IATTC staff met with fishers to develop and disseminate procedures and technologies to reduce incidental mortality. New technologies developed by fishers, such as panels of finer-mesh net to help dolphins escape ("Medina panels"), were disseminated in these workshops.[24]

Thereafter, under IATTC auspices, progressively stricter limits on dolphin bycatch were created. In 1992 participants in the fishery reached agreement on a multilateral basis to reduce dolphin mortality, the La Jolla Agreement. The main provisions were passed as a resolution at the IATTC's annual meeting in June 1992. Outside the auspices of the Commission, a nearly identical agreement was accepted by states including Colombia, Ecuador, Mexico, and Spain (not then members of the IATTC) (Bayliff 2001, 40).

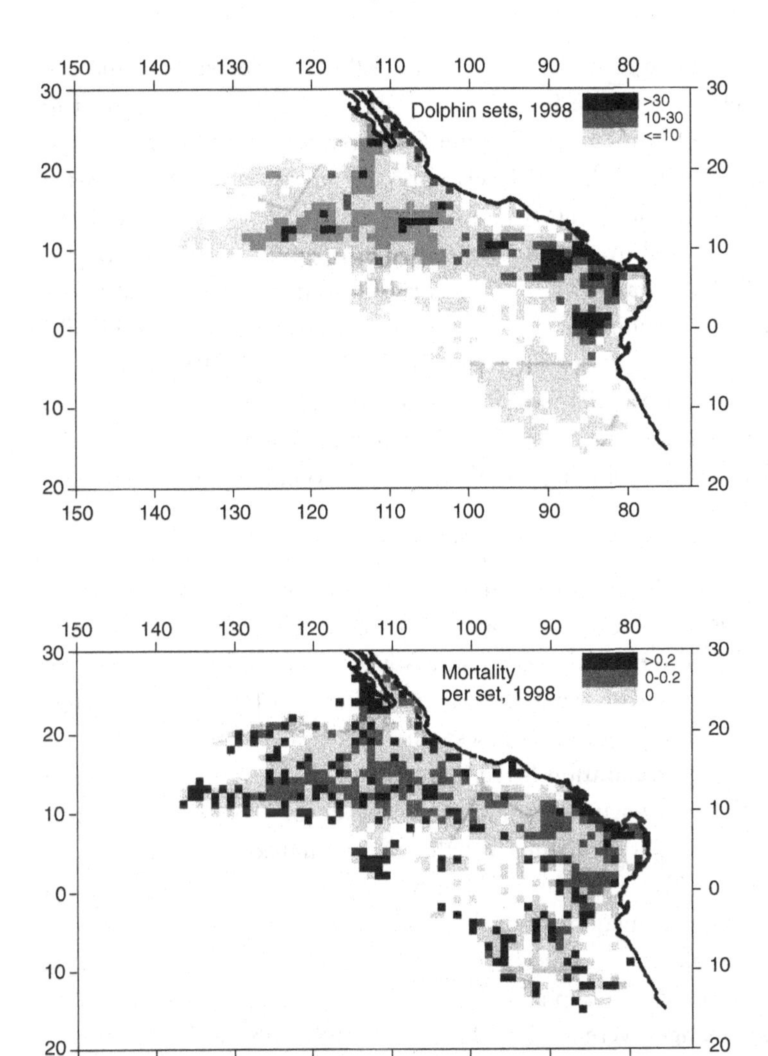

Figure 6.4
Dolphin Sets, 1998. Spatial distributions of the numbers of dolphin sets (upper) and of the average mortalities per set for all dolphins combined (lower) during 1998, as determined from data collected by IATTC observers.
Source: IATTC, Annual Report (1998), 154, fig. 76.

The La Jolla Agreement included the following elements. It allowed dolphin sets to continue in the eastern Pacific Ocean. States agreed to implement an International Dolphin Conservation Program that would reduce dolphin mortality to levels approaching zero by 1999. It set a dolphin mortality limit (DML) for each vessel, equal to the total limit for the fishery divided by the number of vessels operating that year (Bayliff 2001, 40). Each class 6 purse seine vessel (ordinarily the type of vessel that fishes on dolphins) would be required to carry an observer. Finally, the agreement created an International Review Panel (IRP), consisting of the IATTC staff, delegates from member governments, firms, and NGOs, to report annually on allegations of noncompliance. The La Jolla Agreement was nonbinding, but the degree of compliance has been quite high (Parker 1999).

By 1995 vessels fishing on dolphins had reduced dolphin bycatch substantially, but U.S. sanctions remained in place. In 1995 participants in the fishery issued the Declaration of Panama. It established a permanent cap on dolphin bycatch on a stock-by-stock basis (0.2 percent of minimum estimated abundance through 2000, and 0.1 percent thereafter) and called for a program to assess and reduce bycatch in the fishery. Annex I to the Declaration listed "envisioned changes in U.S. law" as a result of this agreement, including (1) lifting of the U.S. embargoes (primary and secondary) on tuna caught in compliance with the La Jolla Agreement and the Panama Declaration; (2) provision of market access for tuna; and (3) amendment of the definition of "dolphin safe" for purposes of compliance. The Declaration of Panama reflects a precautionary approach to reducing dolphin bycatch.[25]

Further changes were made in February 1998, with the Agreement on the International Dolphin Conservation Program (AIDCP). The agreement superseded the La Jolla Agreement and is considered binding. Further, the AIDCP established mortality limits for individual stocks of dolphins (the La Jolla Agreement combined different stocks into a single limit); provides for reduction of bycatch and discards of juvenile fish; requires states to certify to the IATTC that they comply with its provisions for dolphin protection, and that vessels subject to their jurisdiction carry observers; and establishes a system for tracking and verifying that tuna have been caught without injuring dolphins (Bayliff 2001, 41). In short, in the 1990s, the IATTC made significant changes in formal and informal rules to reduce dolphin bycatch and to extend protection to other nontarget species and to juvenile tuna.

According to Parker (1999), "[The IATTC's regulatory program is] one of the most innovative and effective environmental regimes in the world. . . . [It] has reduced dolphin mortality by over 99 percent while eliciting a very high level of compliance from all fishers and flag states."

The history reviewed in this section suggests that the types of knowledge generated by the IATTC as well as its use of this knowledge to ground its regulatory recommendations provide prima facie evidence for the operation of the statutory fix and the committee fix in knowledge generation.

6.4 Limitations of the Interest Group Approach as Applied to Yellowfin Regulation

The interest group approach does not explain the variation in the IATTC's knowledge about the status of yellowfin. From the late 1950s until 1960, the IATTC accepted the belief that effort directed at yellowfin was below the estimated point of MSY. The Commission's regulatory goal, as stated in Article II (5) of the Convention, was to promote fishing effort that maximized sustainable yields. Given this belief and this goal, before 1960 the Commission decided not to regulate.[26] The belief and the action were consistent with *individual firms'* interest in maximizing net annual revenue.

However, each year from the 1960s through the 1970s (also in the late 1990s), the IATTC accepted the belief that effort directed at yellowfin stocks was above the estimated point of MSY.[27] Given the Commission's goal (which did not change), it decided to regulate by use of a global quota (see table 6.1). Assuming that states' shares of the overall quota did not change, regulation was *contrary to* firms' interests each year that the IATTC did not increase the quota or reduced the quota. This was the case each year between 1974 and 1982.

Limitations of Analysis Based on Firms' Collective Interest

It is possible to view regulation (after 1966) as action the IATTC took in the firms' *collective interest.* In other words, it was in the interest of individual firms to maximize net annual income by increasing catches (and effort). Once total effort directed at the fishery moved beyond the

MSY point, however, the collective interest of the firms was to reduce total catch (and effort) to restore abundance. After 1960 overfishing in the eastern Pacific can be seen as a failure of collective action, captured by Prisoner's Dilemma and "tragedy of the commons" models (Olson 1965; Hardin 1968; Axelrod and Keohane 1986). Stressed fisheries are a classic case, in which some form of governance is needed to attain the collective interest in reducing effort to restore abundance (FAO 1996).

However, analysis of the collective interest of firms does not account for the change in the IATTC's beliefs. The staff found that effort directed at yellowfin prior to 1960 was below MSY. Between 1961 and the late 1970s, the staff found effort directed at yellowfin to be at or beyond MSY. The collective interest of the firms—to maximize net annual income—was constant. Therefore, the change in the IATTC's beliefs about stocks cannot be merely a reflection of the firms' collective interest.

Further, the change in the IATTC's belief occurred approximately a decade before NGOs became regularly involved in the organization's activities. NGOs were not actively involved in IATTC matters in the 1950s and early 1960s, when the shift in knowledge occurred. Environmental NGOs were not concerned with the tuna fishery in the eastern Pacific before the early 1970s (Parker 1999). The point is not that interests "don't matter." Rather, an interest-based explanation does not account for the variation in beliefs about yellowfin from the 1950s to the 1990s.

The Importance of Institutional Mechanisms

Crucially, it was the condition of the yellowfin that varied, and the IATTC's beliefs reflected these changes. Using the framework of conservation, which was embedded in the Convention, the Director and the staff developed models to estimate MSY, YPR, and other indices of abundance (the statutory fix). The IATTC's scientific staff used these models to generate data on the status of stocks, which varied year by year, as shown in figures 6.1 and 6.2. Each year, the IATTC formally accepted some of these statements and published them in its Annual Report (the committee fix). Shaped by institutional rules and practices, the IATTC generated knowledge about the status of yellowfin, which varied year by year.

Over time, changes in the IATTC's knowledge shaped changes in its regulatory actions. The firms' collective interest was to maximize yields from the fishery. This interest, assumed to be constant over time, cannot explain the observed variation in the IATTC's regulatory actions (table 6.1). Furthermore, the change in knowledge and regulatory action occurred more than ten years prior to the involvement of environmental NGOs. Once again, interests of firms (which did not change) or NGOs (not yet represented) cannot explain the change in regulatory action between 1950 and the late 1970s.

Change in the IATTC's beliefs can account for the change in regulatory action in this period. Agents needed to form some beliefs about the status of stocks before they could decide what action was in their collective interest. If stocks were abundant, it would be in the firms' collective interest to increase fishing. If stocks were depleted, it would be in the firms' collective interest to limit or to decrease fishing. As a result of changes in beliefs about yellowfin stocks, the IATTC's recommendations for regulatory action varied.

6.5 Limitations of the Interest Group Approach as Applied to Dolphin Regulation

The interest group approach by itself also cannot explain how the policy preferences of firms and NGOs changed as a result of new knowledge generated by the IATTC's scientific staff. Firms' policy preferences changed as a direct result of knowledge produced by the staff. For example, the staff's findings persuaded vessel captains that incidental mortality in the fishery was very high, consistent with the complaints of NGOs (Parker 1999, 63). In addition, working with the staff, the captains accepted certain new technologies and procedures that, at minimal additional cost, greatly reduced incidental mortality by the 1990s.[28]

Knowledge produced by the IATTC's scientific staff also changed the policy preferences of some NGOs. In the mid-1990s, the staff produced a study of the ecological impact of alternative methods of fishing. There are three methods involving purse seine gear: (1) setting purse seine nets on dolphins (dolphin fishing); (2) setting nets on schools of fish; and (3) setting nets on logs. Compared to dolphin fishing, setting nets on schools catches a higher proportion of juvenile fish. Setting nets on logs (or other floating objects) tends to ensnare various nontarget species, including sea turtles. The study concluded that fishing on dolphins, using appropriate gear and methods to release them before hauling in the net, is less harmful to the ecosystem than alternative methods (Hall 1998). This analysis shifted the attention of many NGOs from dolphin protection to ecosystem protection.[29] That is, in this case, a change in knowledge led to a change in the preferences or interests of both firms and NGOs.

6.6 Summary

Changes in knowledge about phenomena (the abundance of tuna stocks and the level of dolphin mortality) drove changes in the IATTC's regulatory actions from 1950 to the present. Beliefs (or ideas) were not merely "hooks" that agents used to justify self-interested actions. The present chapter explained how the acceptance of particular beliefs is necessary *before* agents can decide what action is consistent with their interests.

Furthermore, the claim is not that beliefs *rather than* interests matter. Much of the regime literature assumes that ideas explain some residual variance in policy outcomes (Hasenclever, Mayer, and Rittberger 1996). The value added by the present chapter is to elaborate the *institutional mechanisms* through which agents generate and accept particular beliefs about phenomena (like fish or dolphins). Once accepted, beliefs (e.g., stocks are abundant/depleted) clarify *which* action (e.g., no regulation/regulation) is consistent with agents' interests.

This chapter has argued that the statutory fix and the committee fix are mechanisms through which the IATTC formed beliefs about the status of yellowfin and dolphin stocks. In the early 1950s, the Director had planned a more ambitious, ecosystems-based program. Faced with budget cuts, he scaled his plans back, prioritizing work based on the goal of the Convention, to promote conservation. The embedded framework shaped subsequent research under IATTC auspices (statutory fix). From its earliest meetings, the IATTC developed standard procedures for accepting scientific statements (committee fix), and translating science advice into policy. Whereas interest groups—particularly fishing firms—might have generated their own findings by hiring experts, this did not happen. By creating institutionalized procedures for translating science advice into group beliefs, the IATTC successfully avoided public deconstruction of its scientific findings by interest groups.

7

Conclusion and Paths for Future Research

Some of the world's most pressing environmental problems, like climate change and loss of species, have an important human dimension. Human beliefs about environmental problems—whether they exist, how severe they are—shape human action. To understand social action, therefore, we must develop some explanation as to how people, either individually or in groups, form beliefs. To date, most of the international relations literature has focused on how ideas, especially new scientific knowledge, affect the formation and functioning of institutions (Litfin 1994; Haas 1990; 1992; 1998). Much less attention has focused on the prior influence of institutions on the generation of new knowledge. In short, it is necessary to turn the causal arrow around, to study how institutions shape the generation of new knowledge.

7.1 Recapping the Argument: Institutional Dimensions of Knowledge Generation

In 1877 the philosopher Charles S. Peirce wrote, "Unless we make ourselves hermits, we shall necessarily influence each others' opinions; *so that the problem becomes how to fix belief, not in the individual merely, but in the community*" (13) [emphasis added]. In emergency situations, where there is uncertainty about facts related to a particular threat, people must form beliefs before deciding what action to take. ("Where is the fire coming from, and how fast is it moving?" precedes "Should we run that way?") This happens in more routine situations as well, for instance, a group of central bankers must decide what the current rate of inflation is before deciding on monetary policy. In general, the problem is how groups generate knowledge or fix beliefs. The argument is that institutions shape knowledge through the following mechanisms: the positional fix, the statutory fix, and the committee fix.

To explain the institutional dimensions of knowledge generation, chapter 2 laid some conceptual groundwork. Complex political, economic, and epistemic systems have an institutional dimension. They consist of clusters of social roles, to which rights and rules (or responsibilities) are attached. When people act as group members (qua group members), they refer to the rights and rules attached to their roles when deciding what to do. Secondary roles like great power or hegemon define the relationship between a group and other groups. Primary roles like scientist or NRC member define the relationship of an individual to a group.

Use of a role position to generate knowledge or to form beliefs is an example of the *positional fix*. Chapters 3–5 provided empirical support for the claim that global political changes are translated into changes in knowledge through individuals like William Ritter or Roger Revelle, who simultaneously occupied the roles of scientist (Director of the Scripps Institution) and political adviser (member of NRC, DOD committees, etc.). The role position was the mechanism through which global institutions (e.g., great power status) shaped the generation of new knowledge (the research program at the Scripps Institution).

People use ideas, embedded in rules, when deciding what kinds of knowledge to generate. Chapter 6 provided evidence that at the IATTC, Director of Investigations Milner B. Schaefer could have developed ecosystems models in the 1950s, and he considered doing so. However, he referred to the Convention for the Creation of an Inter-American Tropical Tuna Commission (Article II), and decided that the purpose of his scientific staff had been clearly stated in the statute. The purpose was to estimate maximum sustainable yield to guide fishery regulation each year. The framework, MSY, which was embedded in the statute, shaped the Director's decision to have his staff develop MSY, YPR, production, and other models. It shaped the kinds of data the IATTC scientific staff generated. This is an example of the *statutory fix*.

Many organizations develop standardized procedures for accepting beliefs as groups. Again, chapter 6 provided evidence using the example of the IATTC. Each year, the IATTC has held a meeting involving some of the scientific staff and the Commissioners from contracting states. At the meeting, the Director of Investigations provides a summary of the work completed by the scientific staff. This includes, for example, summaries of abundance of species including yellowfin. When accepted by the Commission in plenary session, and published as an Annual Report, these data constitute the group's beliefs about the status of the fishery. Such regularized procedures are likely to be very common, for instance, central bankers must decide what the inflation rate is before they set interest rates, security experts must clarify the nature of a threat before setting defense policy, environmental managers must decide whether an ecosystem is threatened (and how) before recommending regulations. When regularized, when there are standard rules and procedures for arriving at a group belief about matters of fact, this is an example of the *committee fix*.

The empirical chapters (3–6) explored two alternative explanations, neorealism and the interest group approach. It is necessary to note that work like Litfin's (1994), which was among the first to emphasize the importance of knowledge in global environmental politics, is not considered an alternative explanation. This is due to differences in foundational assumptions (see chapter 1). While valuable as a source of thought-provoking criticism, poststructural analysis tends to privilege different ontological phenomena, to ask different kinds of questions, and to pursue different intellectual goals. Because poststructuralism is incommensurable with institutionalism, the empirical chapters do not take it on as a rival explanation.

The empirical chapters compare the predictions of institutionalism with those of neorealism and the interest group approach. In brief, the institutional approach predicts that the Scripps Institution would generate defense-related research when the United States accepted a role as a great power but not otherwise (the positional fix). The United States accepted the role of great power during World War I but not before (1900–1916) or after (1920s and 1930s) (Carr 1939; Wight, in Toynbee 1952; Hillman 1952; Bull 1977, 212; Kennedy 1988, 328; Gaddis 1997, 6–7). The neorealist approach predicts that the Scripps Institution would generate defense-related research when the United States ranked as a hegemon, based on its global rank in terms of war-fighting capabilities (Waltz 1979). The United States had the capabilities of a regional hegemon in the Western Hemisphere from about 1890. Chapters 3–5 provide evidence to support the institutional argument, not the neorealist one.

Chapter 6 provided evidence that institutional mechanisms, not interest groups, shaped the IATTC's research and regulatory activities. Ideas embedded in rules account for Schaefer's selection of MSY as a guide to research (the statutory fix). Each year, the IATTC generated and selected certain statements about the abundance of yellowfin and published them in its Annual Reports (the committee fix). As the status of the stocks changed, from abundant in the 1950s to increasingly depleted in the 1960s and 1970s, the IATTC's beliefs and regulatory actions changed. The interests of the fishing firms, to maximize profits (individually or collectively), did not change from the 1950s to the 1970s. Nor were NGOs active at this time. Therefore, the activity of interest groups, which did not change, cannot explain the observed change in the IATTC's beliefs or in its increasingly strict regulation of yellowfin (until 1979). In short, the institutional approach explains the generation of new knowledge and the acceptance of beliefs better than potential alternatives.

7.2 Institutions, Knowledge, and Action

Students of international relations commonly bifurcate the effects of power from the effects of institutions. It has been common for them to argue over whether the international power structure or institutions matter more (Keohane, ed., 1986). By the 1980s, neorealists and regime theorists seemed to agree that both power and rules mattered, each explaining some part of the variance in international outcomes (Krasner, ed., 1983; Keohane 1988). By the early 1990s, many regime theorists also added "ideas" to the mix of explanatory factors that "mattered" (Goldstein and Keohane 1993; Hasenclever et al. 1996).

Taking a different tack, inspired by classic works such as Bull (1977) as well as by social practice approaches (Tuomela 1995; 2000a; 2000b; 2002), this book argues that power itself has an institutional dimension. Roles that states may acquire, like sovereign or hegemon, entail the threat or actual use of force. Further, these political institutions tend to be generative of new knowledge. Through certain mechanisms—hybrid roles like scientist/political adviser—political purposes are injected into research agendas. Power and knowledge are deeply intertwined (Litfin 1994). At the same time, it is important to identify causal connections between the two, where possible.

This book builds on work by other institutionalists, extending it to study the interrelations of power and knowledge. Before the United States accepted a role as a great power, the Scripps Institution's research agenda focused primarily on marine life. The Director's role was to coordinate research on chemical, physical, and biological aspects of plankton ecology, in a program modeled after the Kiel school in Germany.

As the United States emerged as a great power during the First World War, the Director's role acquired a political dimension. As a member of the NRC, he was to consider how the Scripps Institution's research could further the war effort. During the postwar isolation, from about 1920 to 1939, the United States abandoned its global role. Consequently, research related to national defense disappeared from the agenda. When the United States emerged as a global hegemon after the Second World War, Directors of the Scripps Institution again participated in hybrid networks, including the NRC and various advisory committees of the Departments of Defense and Energy.

The point is that global political changes, seen here as institutional, drove changes in the generation of new knowledge at the Scripps Institution. When the United States acquired its hegemonic role, Scripps acquired a new, global outlook. New lines of research appeared, including research on the global carbon cycle. It was the change in global political institutions that shaped the broad outlines of postwar science at the Scripps Institution. U.S. hegemony entailed responsibilities, like defending the entire Pacific basin, that generated new lines of research.

Work like Charles Keeling's measurements of carbon dioxide in sea water was pioneering in 1957. The Scripps Institution helped to set off some of the earliest alarms about global climate change. Climate change research programs grew in subsequent decades (Social Learning Group 2001). By 1995 scientists around the world reached a consensus that global climate change is occurring (Houghton et al. 1995). By 1997 diplomats had reached agreement on a plan to reduce emissions of greenhouse gases (Victor 2001). New technologies, such as lower-emission hybrid cars, began to reach the market. Although the prospects for the Kyoto Protocol are uncertain, the issue has remained on the global agenda for more than a decade.

The present study shows how it is possible to acknowledge that institutions are social constructions without giving up on the possibility of identifying causal properties of institutions. Social institutions, once constituted, generate patterned behavior. In the present study I have looked at the systematic effects that institutions can have on knowledge generation, and the mechanisms through which they have these effects. Acknowledging both the social construction of institutions and the possibility of analyzing the systematic patterns they generate, and especially the types of knowledge they generate, is important for a number of reasons. First, it encourages us to question whether current institutions adequately support or prevent generation of knowledge that is needed to solve various problems we face. Second, it encourages us to consider what the knowledge-generating properties of alternative institutional arrangements might be. Third, attention to the mechanisms through which institutions generate knowledge might lead to a better understanding of how the knowledge-generating properties of given institutions might best be harnessed. The present study has identified one set of mechanisms through which institutions generate knowledge; it remains for future research to determine how general the mechanisms investigated are, and how they interact with other mechanisms.[1]

7.3 Institutional Dimensions of Scientific Research

Whereas the Scripps Institution illustrates the effects of global political changes on the generation of new knowledge, its neighbor, the IATTC, provides a clearer view of how interests matter. From the beginning, the IATTC functioned as a hybrid organization, generating scientific research to support the operations of a commercial fleet.

Analysis of research results, particularly on the status of yellowfin from the 1950s to the 1970s, does not support the view that interest group activity determined the content of scientific research. Nor does standard regime theory account for this change in the IATTC's beliefs. Following Krasner (1983), one might predict that since the international power structure was constant during the period under consideration, regime rules would matter more. But since the regime rules—the purposes and function of the IATTC—did not change either, standard regime theory cannot account for the change in the IATTC's beliefs or actions. The IATTC's beliefs changed, driving the change in its recommended regulations. But how?

To answer this question requires an extension of standard regime theory, to explain how institutions enable an organization like the IATTC to form beliefs about phenomena like the status of yellowfin. The Convention that created the IATTC embedded certain ideas, like MSY, that shaped the content of research. The Director used the Convention as a guide when he designed the early research program, in the 1950s. Furthermore, the IATTC created standard procedures, like an annual scientific meeting, that facilitated the formation of group beliefs about the status of stocks. The institution was designed to allow the organization, each year, to fix a new belief about the status of stocks, in order to guide decision making. The further study of such knowledge-generating mechanisms promises to be rich terrain for future research.

The knowledge-generating function of institutions also suggests that we need not assume that gaps in knowledge should always be understood as failures of collective epistemic action. In at least some cases, the lack of relevant knowledge may have institutional sources rather than a simple lack of incentive to individuals to produce the relevant knowledge. It remains a question for future research to determine which problems might be fruitfully characterized in this manner.[2]

7.4 Paths for Future Research

The interrelationships between institutions and knowledge provide at least five avenues for future research. First, more work is needed on how institutional rules shape market incentives, which in turn influence the generation of new technologies. For example, it has been widely noted that the Montreal Protocol successfully stimulated investment by DuPont and others in substitutes for ozone-depleting chemicals (Benedick 1998; Makhijani and Gurney 1995). However, some critics argue that the Kyoto Protocol was poorly designed and is unlikely to have a similar effect (Victor 2001). More research is needed on how to best design institutions to provide incentives to firms and individuals to generate the kinds of clean technologies societies require.

Second, work on failures of collective epistemic action has so far only begun to scratch the surface. Extending earlier work by Mitchell (1994; 1998), Chayes and Chayes (1990; 1995), and Hønneland (2000), more research is needed on how institutions can be designed to improve transparency. Lack of transparency is associated with some of the more severe cases of noncompliance, like the near-elimination of the great whales in the 1950s and 1960s (Walsh 1999). Critics argue that unless mechanisms are designed to improve transparency of the emissions trading system, the Kyoto Protocol will be threatened (Young, ed., 1999; Victor 2001).

Third, more research is needed on how dispute settlement mechanisms shape the generation of knowledge. Typically, litigants provide evidence for a particular point of view. The self-interest of litigants helps to ensure that all relevant evidence comes to light (Arrow 1995). At the same time, judicial rulings can be seen as a group belief. Some rulings, such as the decision in 1999 by the International Criminal Tribunal for the former Yugoslavia to indict Slobodan Milosevic, may serve as precedents and guide future decisions (Goldstein et al. 2000, 385). In this sense, the decisions taken by dispute settlement panels may serve as precedents, guiding the path of future decisions. Future research could explore any such effects of international dispute settlement on the evolution of legal decision making.

Fourth, more work is needed on the mechanisms through which institutions generate and sustain social facts, for example, by perpetuating reflexively held beliefs. Institutions may generate reflexively held beliefs, for example, that paper counts as currency in certain contexts. They may also play a part in sustaining social facts like trust and legitimacy. Where institutions fail to generate such beliefs, these social facts and associated practices could disintegrate.

Fifth, more research is needed on the institutionalized networks, often built into environmental regimes, for the generation of scientific knowledge. These networks often take the form of scientific advisory committees like the National Research Council, which bring scientists and policymakers together. Where political and scientific institutions overlap, there are often built-in contradictions. For example, scientists may prefer to insulate their research from the supposed taint of political influence, to safeguard the perceived integrity of their work (Andresen et al. 2000). However, to design public policy for problems such as global climate change, societies need research results that have clear implications for policy. In such situations, institutional design is particularly important.

7.5 Conclusion

People are often forced to make important decisions under conditions of uncertainty. In such situations, they must commit to a particular belief—although they cannot be sure that it is true—in order to define their interests and decide what action to take. For example, international environmental organizations must reach some consensus on the existence of an anthropogenic threat, say, to the climate system, before they can meaningfully consider efforts to change deeply engrained patterns of human behavior. Groups like the IATTC must decide whether there is a threat to marine ecosystems (or parts of marine ecosystems) before they decide whether to recommend regulations on commercial fishing in the eastern Pacific Ocean.

To date, most of the international relations literature has *not* focused on the question of how institutions shape knowledge or how organizations (like the IPCC or the IATTC) form beliefs. Rather, most of this literature to date has focused on how new knowledge—scientific discoveries or epistemic consensus—shapes international cooperation (Haas 1990; 1992; Litfin 1994). But it is necessary to turn the question around, to explore how complex institutional systems shape the generation of knowledge. Using evidence from the history of research at the Scripps Institution and the IATTC (roughly from 1905 to the late 1990s), this book has argued that institutions shape knowledge through the following mechanisms: the positional fix, the statutory fix, and the committee fix.

Human actions are now important drivers of environmental change like climate change and species extinction (Houghton et al. 1995; NRC 1995). The beliefs people accept ("what people know") about these problems shape the way they perceive their interests, and how they act. To the extent that institutions shape people's knowledge about phenomena like climate change or marine ecosystems, it is important to better understand *how* (that is, the mechanisms through which) this happens. By highlighting specific mechanisms or "fixes" (positional, statutory, committee) and by outlining a broader agenda for future research, this book opens a path toward clearer understanding of the knowledge-generating function of institutions.

Notes

Chapter 1

1. The structured, focused case study in chapter 6 shows how changes in knowledge about the abundance of tuna and dolphin stocks reshaped international regulatory action to protect them.

2. The assumptions are explained in more detail in section 1.5.

3. The distinction between weak and strong constructivism is Ruggie's (1998). By strong constructivism, I mean poststructuralism and other versions of postmodernism. For readers interested in finer distinctions, section 1.5 compares my foundational assumptions with other approaches.

4. For a discussion of "collective action" and "social practice" versions of institutionalism, see Young (2001). For a classic treatment of rational choice vs. reflective (or constructivist) approaches, see Keohane (1988).

5. The term is borrowed from Searle (1995). "Brute facts" refers to phenomena that would in principle exist even if human beings did not, like rocks or trees. "Social facts" refers to phenomena that would disappear if human beings did, e.g., the use of specially printed paper as money.

6. The phrase "repair uncertainty" is borrowed from Jasanoff (1997, 232).

7. Raimo Tuomela, personal communication, March 2002. For a discussion of how group beliefs can be justified, see Tuomela (2002).

8. Chapter 6, on the IATTC, illustrates this mechanism.

9. As explained in section 1.5, the interest group approach is equivalent to neoclassical economic analysis as applied to politics. See, for instance, Olson (1965); Caporaso and Levine (1992, ch. 6); Moravcsik (1997).

10. For an introduction to the new institutionalism in the social sciences, see Young (1994).

11. See also Keohane (1988). Keohane divides institutionalists into rationalists and reflectivists. Many institutionalists who reject economic approaches self-identify as constructivists.

12. Following Turner (1985, 79), I use the term *identity* to refer to collective awareness, in the sense that people perceive themselves to be a distinct social

unit. Identity is therefore similar to what Tuomela (1995) calls "we-ness." Social roles are more formalized, in the sense that particular clusters of tasks and rights become associated with individuals acting in particular social positions. When a stabilized set of tasks and rights becomes associated with a social position, the position (and associated tasks and rights) is a *social role*. Roles are institutional, but identities are not.

13. See King, Keohane, and Verba (1994, 43). The thickest of case studies abstracts away from conceptual detail. In terms of capturing the "real world," the distance between a formal model and the thickest of descriptions is smaller than the difference between the thickest of descriptions and the real world it attempts to capture.

14. Moravcsik's (1997) "liberal IR theory" is an excellent example of what I call the interest group approach. Caparaso and Levine (1992) review economic approaches to politics. Within economic approaches to politics, neoclassical approaches (which assume interests are fixed) correspond to what I call the interest group approach. An excellent example of thin—more individualist—institutionalism is Milner (1997). The new institutionalism also includes thicker—more social—approaches. For a systematic treatment of social action, see Tuomela (1995; 2000a). On neorealism as an approach to international relations, see, for example, Waltz (1979); Keohane, ed. (1986); Mearsheimer (1994).

15. That is, stems from different, and incompatible, ontological assumptions.

16. Therefore the case studies compare institutionalism with interest group and neorealist alternatives but not with poststructuralism, which is incommensurable.

17. More specifically, Waltz (1979, 100–101) argues for a three-part definition of structure. First, structures are defined according to the principle by which the system is ordered. In international affairs, this principle is anarchy. Second, he argues that structures are defined by the specification of functions assigned to different units. This "drops out," he argues, since the international system is composed of states which (he assumes) do not assign functions or tasks to each other. Third, structures are defined by the distribution of capabilities across units. This third element receives most weight in neorealist theory. In neorealist theory, international structure, set against a context of anarchy, consists of a hierarchy of states based on material capabilities. Powerful states (measured in terms of physical capabilities) coerce and (if possible) eliminate weaker ones.

18. A classic Prisoner's Dilemma model is a two-person game in which each player has the option to cooperate (C) or to defect (D), and in which payoffs are assigned such that, for each player, the preferred outcomes are DC > CC > CD > DD. Since the dominant strategy for each player is the same (D), the joint outcome (DD) is suboptimal for the collective (Axelrod and Keohane 1986).

19. The term *mediated realism* is borrowed from Jasanoff (1997, 231), and this section has been informed by her chapter.

20. Noam Chomsky (2000) argues that the rules of grammar are generative, in the sense that if a person knows a particular rule, she will use it to guide the construction of sentences in new situations. For example, if a person knows that in English one constructs the plural by adding "s", one is likely to try this when

using an unfamiliar word. The strategy does not always work (e.g., fungus/fungi), but it nevertheless guides speech in a wide variety of contexts.

Chapter 2

1. My use of the term *social role* differs from Tuomela (1995, 33), for whom role behavior is connected to social norms in a group but not to formal or informal rules.

2. Tuomela (1995, 180, 195) considers a functioning social group to be a group in which (1) individuals self-regard as group members (a "core group"), and (2) the group has an authority system. An authority system is used by the members of the group to create a group will and a shared group intention, with specific contents and accompanying group commitments. Agents in global affairs—states, firms, nongovernmental organizations, sometimes also individuals—do not qualify as a group in this sense. I use the term *collective* to refer to this heterogeneity of actors.

3. For alternative institutional arrangements that privilege the nonscientific voices in local communities, see Fischer (2000). Johannes (1981) is a classic treatment of local community knowledge of marine ecosystems, which is superior to scientific knowledge. For a fishery management scheme that makes use of local knowledge, see Tjetjep Nurasa et al. "The Role of the Panglima Laot 'Sea Commander' System in Coastal Fisheries Management in Aceh, Indonesia," Regional Office for Asia and the Pacific, Food and Agriculture Organization of the United Nations, 1994 (RAPA Publication 1994-8).

4. Chapters 3, 4, and 5 develop this example more fully.

5. Chapter 6 develops this argument in more detail.

6. There is preliminary evidence in recently declassified archival sources to support the argument that the groups in the national security establishment form beliefs about particular threats (e.g., the threat posed by the USSR early in the Cold War) to justify particular policy actions or strategies. See, for example, Declassified U.S. Government Documents: Letter, Hull, to President Eisenhower, October 30, 1958; Memorandum for the Secretary of Defense from Arthur Radford, May 22, 1956; White House, Operations Coordinating Board, May 18, 1956.

7. As explained in section 1.5, poststructural approaches are incommensurable with institutionalism. Therefore it is not possible to sort out the differences with evidence from case studies.

8. For a neorealist critique of neorealism's lack of attention to beliefs, see Kim and Bueno de Mesquita (1995); Downs, Rocke, and Siverson (1986, 136). Exceptions to resistance to the study of beliefs include Jervis (1976) and Falkowski (1979).

9. Although I have not found these arguments made in published journals, a number of reviewers have responded in this way to earlier drafts of this manuscript, and to parts of it.

10. Again, the discussion of what realists would say cannot be based on published papers because neorealists have not taken up this question. My interpretation of what neorealists would say is based on reviews of earlier drafts of this manuscript.

11. I thank Dave Guston for the "piper" metaphor. The interest group approach in international relations appears in many forms, most prominently in the liberalism of Moravczik (1997). Again, Moravczik does not explicitly take up problems of knowledge, and other versions of economics applied to politics exist (Caparaso and Levine 1990). Therefore, I frame the interest group approach in a generic way.

12. On the importance of using hard cases (a case least likely, on a priori grounds, to accord with theoretical predictions but which does nevertheless) and easy cases (a case most likely, on a priori grounds, to accord with theoretical predictions but which does not), see Eckstein (1975); King, Keohane, and Verba (1994, 209–212).

13. On conservation, see Nash (1982); on MSY as applied to fisheries, see Smith (1994).

Chapter 3

1. To be precise, until 1912 the organization that conducted marine biological work under Ritter's direction was known as the San Diego Marine Biological Station. However, I will refer to the organization as the Scripps Institution and to 1905–1915 as the early years of research at the Scripps Institution of Oceanography. Although historically imprecise, it is consistent with common usage (e.g., Raitt and Moulton 1967; Shor 1978; Mills 1993).

2. Consistent with the distinctions made in chapter 2, *secondary role* refers to relationships among groups, e.g., to the U.S. role in the world (great power, hegemon). Hereafter, "U.S. role" implies "secondary role."

3. For example, according to the historian Paul Kennedy (1988, 248), "The U.S. had definitely become a great power. But it was not part of the great power system. . . . No one was in favor of abandoning the existing state of very comfortable isolation."

4. Sprout and Sprout (1942, 289).

5. Kofoid wrote, "I am sure it would be a very great pleasure to you and perhaps to our donors to hear how universally our station is known and how widely your plans for it meet with hearty approval. The idea of a truly marine biological station with coordinated work directed toward a common end . . . meets with universal approval. Our freedom from "economic" pressure, our unique situation, geographically, climatically, and strategically in the world as a whole, is quickly and enthusiastically appreciated." SIO Archives, Letter, Kofoid to Ritter, September 29, 1908.

6. To avoid cumbersome diction, when referring to a role occupied by an individual (e.g., scientist, director, chairman), I simply use the term *role*. (To be more

precise, I should use the term *primary social role*). Consistent with the distinctions made in chapter 2, "primary social role" refers to a cluster of rights and responsibilities that position an individual with respect to a group.

7. Raitt and Moulton (1967, 33).

8. On the ties between Professor Joe LeConte and John Muir, see Gifford (1996); Cohen (1988). Unlike LeConte, Ritter accepted Darwin's evolutionary theory and rejected vitalism. For a discussion of the influence of LeConte and Muir on Ritter's thought, see Mills (1993, 23–24).

9. SIO Archives, Letter, Ritter to his Uncle Nelson, June 12, 1898.

10. Bancroft Library, Diary entry, July 2, 1899.

11. Muir to the "Big Four Girls," Mary and Cornelia Harriman, Elizabeth Averell, and Dorothea Draper, in Gifford (1996, 338).

12. Bancroft Library, Letter, Muir to Ritter, October 20, 1899. Charles A. Keeler, Captain P. A. Doran (captain of the *Elder,* which carried the Expedition in 1899), C. Hart Merriam (in 1899, Chief of the Biological Survey, U.S. Department of Agriculture), the geographer Henry Gannett, and Professor Charles Gilbert (see Gifford 1996, 249, 333, 878–880, 889; Cohen 1988, 12; William Ritter 1899, 226; Mary Ritter 1933; Raitt and Moulton 1967, 6; see also Muir's 1912 essay "Edward Henry Harriman," in Gifford 1996, in which he describes C. Hart Merriam's role in bringing the group together).

13. Mills (1993, 22–24) points to LeConte as an important influence on Ritter. It is important not to overstate the innocence of Ritter's early thought: some of his views were ugly. Both men were racist; LeConte as a traditional Southerner, and Ritter in his tendency to view Asian immigrants in California as a "peril" to whites.

14. John Muir, *My First Summer in the Sierra* (1911). This statement was redrafted from Muir's journal entry, July 29, 1869: "When we try to pick out anything by itself, we find that it is bound fast by a thousand invisible cords that cannot be broken to everything in the universe. I fancy I can hear a heart beating in every crystal, in every grain of sand I see a wise plan in the making and shaping and placing of every one of them." (Quoted in Fox 1986.)

15. SIO Archives, Letter, Ritter to his uncle, Ritter Correspondence, 1889–1890. On Agassiz and his lab, see D. J. Zinn, "Alexander Agassiz and the Financial Support of Oceanography in the United States," in Sears and Merriman (1980, 86).

16. Bancroft Library, Letter, Ritter to Agassiz, April 13, 1905; see also Mills (1993, 12); Raitt and Moulton (1967, 36).

17. See also Day (1999). Zoologists like Bigelow at Harvard were not impressed with Ritter's program. Schlee (1973, 231–233).

18. SIO Archives, Report, "Biological Station, 1903"; and William Ritter (1908, 329).

19. SIO Archives, Day, "Scripps Benefactions." On the Stazione, see Jane M. Oppenheimer, "Some Historical Backgrounds for the Establishment of the Stazione Zoologica at Naples," in Sears and Merriman (1980).

20. William Ritter (1911); and Bancroft Library, Letter, Ritter to William James, March 1908.

21. According to Golley (1993), Tansley coined the term in a 1935 article in the scientific magazine *Ecology*.

22. Ritter did write in a general way of nature as a system. See, e.g., William Ritter (1919, 10). Although his organismal philosophy may have described ecological phenomena in systemic terms, his research program did not trace energy flows or carbon flows as more fully developed systems theories did beginning in the 1930s.

23. Mills (1993, 20) makes the connection between Ritter's work and Whitehead's *Science and the Modern World.* Golley (1993) explains the importance of Whitehead and Leopold as forerunners to ecosystems theory. 1–206.

24. SIO Archives, Scripps Institution for Biological Research, Annual Report to the President, 1915, 200.

25. Ritter, Report to the President of the University of California, 1917/1918, 3.

26. Ritter, Report to the President of the University of California, 1923/24, 121.

27. Ritter explained in a report to the University of California dated July 1, 1914, that Sumner's testing of Weisman's hypothesis (that characteristics acquired from an organism's adaptation to its environment are not transmitted to offspring) had bearing on the general program at the Scripps Institution. Ritter believed Weisman's hypothesis had bearing on his work related to marine organisms' adaptation to the marine environment (51, 52).

28. SIO Archives, Scripps Institution for Biological Research, Annual Report to the President, 1919, 11–12. A shorter version of the same book was published as *An Organismal Theory of Consciousness* in 1919.

29. SIO Archives, Ritter, Annual Report of the Director, 1913/1914, 117.

30. For a review of research on submarine warfare at this time, primarily in the UK, see Hackmann (1984).

31. Businessmen from Del Mar, a neighboring community, tried to woo the biological station away from La Jolla. Presumably their commercial interest was to boost property values. See SIO Archives, Letter, Kofoid to Samons Fletcher Investment, August 1906.

32. Bancroft Library, Letter, Ritter to President David P. Barrows, April 19, 1923. See also SIO Archives, Scripps Institution for Biological Research, Annual Report to the President, July 1, 1923, 127–128.

Chapter 4

1. On the U.S. retreat from a global role, see Gaddis (1997, 34, 6–7); Kennedy (1988, 328); Bull (1977, 212); Hillman (1952); Wight, in Toynbee (1952); Carr (1939).

2. Bancroft Library, Letter, Merriam to Ritter, March 31, 1917.

3. Industry figures from SIO Archives, Ritter, Annual Report to the University of California, 1917.

4. SIO Archives, Ritter, Annual Report to the University of California, 1917, 7.

5. Bancroft Library, Letter, Ritter to Merriam of the Carnegie Institution in Washington, August 3, 1922.

6. Bancroft Library, Letter, Ritter to Merriam, October 9, 1922.

7. Bancroft Library, Letter, Ritter to Mrs. Ritter, November 17, 1922.

8. "Help" meant that the Carnegie Institution would fund some projects at the Scripps Institution. See Day (1999, 71).

9. Carnegie supported the work of Vilhelm Bjerknes (consistently since 1905) and had also supported the work of Harald Sverdrup (Nierenberg 1996). On Bjerknes's life and work, see Friedman (1989).

10. Deborah Day notes that Harald Sverdrup, a student of Bjorn Helland Hansen (himself a student of Bjerknes) was in residence at the Carnegie Institution at the time of Ritter's meetings with Merriam, Barrows, and Vaughan. Sverdrup impressed upon the men the importance of *physical* oceanography (Day 1999, 72).

11. As Gaddis (1997) notes, that the United States declined to "maximize its power" in the 1920s and 1930s contradicts a fundamental assumption of neo-realist theory.

12. From Wight (1952).

13. Ritter had promised Sumner in November 1922 that his work would "not be ignored" when the Institution changed from biological to oceanographic. Most likely, Sumner thought his experiments on deer mice would not be excluded from the new program (Bancroft Library, Letter, Ritter to Mrs. Ritter, November 17, 1922).

14. SIO Archives, Letter, Bjerknes to Vaughan, March 28, 1926.

15. SIO Archives, Letter, Vaughan to Sproul, February 6, 1934.

16. SIO Archives, Letter, Vaughan to Louderback, December 1935.

17. Friedman (1994, 29). Nierenberg (1996, 13) adds that Sverdrup thought McEwen's weather forecasts were "close to witchcraft" and terminated them.

18. The University of California made Sverdrup's appointment permanent in 1941. He had planned to return to Norway at the end of that year (Raitt and Moulton (1967, 135).

Chapter 5

1. SIO Archives, Report, Sverdrup, April 25, 1944.

2. According to Bull (1977), hegemony is a situation in which a great power dominates smaller powers in its sphere of influence but does not routinely resort to force as it exercises its influence. I use *hegemony* to refer to an institutional

role that a great power accepts. Gaddis (1997, 13, 49) argues that the United States accepted a role as a great power after World War II and did not shy away from the role as it had after World War I. He further argues that Europeans accepted U.S. hegemony and initiated the creation of the North Atlantic Treaty Organization. Kennedy (1988, 357–361) argues that the U.S. position was founded on preponderant military and economic capabilities.

3. On deterrence, see Bull (1977, 118–126). The relationship developed between the United States and the USSR in the mid-1950s. Bull argues that deterrence functioned such that it preserved peace, prevented war between the United States and the USSR, and preserved the global balance of power. In this sense, maintaining a deterrent force was consistent with the U.S. (secondary) role as a great power, defined in chapter 2.

4. Day (1985, 10–11); SIO Archives, Letter from Solberg to Eckart, March 22, 1948.

5. SIO Archives, Annual Report of the Director, 1938–1942.

6. SIO Archives, Letter, Sverdrup to Sproul, June 15, 1942.

7. For a full discussion of the charges against Sverdrup, as well as the efforts of his colleagues (particularly Roger Revelle) to persuade the Navy to dismiss them, see Oreskes and Rainger (2000).

8. In a report to Deutsch, Comptroller, University of California, Berkeley, dated August 1938, Sverdrup estimated the total annual budget as $87,900 in 1938-39, $86,050 in 1939-40, and $86,850 in 1940-41. Scripps' actual total budget in 1939-40 was $98,500 (SIO Archives, Letter, Sverdrup to Nichols, February 19, 1940). In 1943-44 the total budget was just over $87,000. Sverdrup reported a donation from the Scripps family of $24,000 and a contribution from the State of California of $53,190 (SIO Archives, Letter, Sverdrup to Revelle, July 3, 1943).

9. SIO Archives, Letter, Sverdrup to Revelle, July 20, 1943.

10. Oral history interview, Calvert (1976a, 4).

11. See, for example, Chayes and Chayes (1995) on arms control agreements. During the Cold War, the United States and the USSR accepted and largely complied with agreements not to expand antimissile defenses, and to manage the growth of their nuclear arsenals.

12. "Secondary role" here refers to the U.S. role as great power or hegemon, competing with the USSR and leading a block of allied states, primarily Western Europe and Japan.

13. SIO Archives: Day, "Introduction," 12; Haggett, June 28, 1946, Revelle Papers, Box 1, f31, "Navy Service, 1946–1947."

14. Hackmann (1986, 109); SIO Archives: Letter, Revelle to Harrington, December 29, 1950; Memorandum, Revelle to Knudsen, July 3, 1952.

15. Day (1985, 11); John Isaacs, To an Era, *Bear Facts,* October 1964, 1.

16. Walter Munk Papers, SIO Archives, 82-57, Box 2, folder entitled "Bikini 1946, General Oceanographic Charts and Reports," quoted in Mukerji (1989, 50).

17. Walter Munk Papers, SIO Archives, 82-57, Box 2, folder entitled "Bikini 1946, General Oceanographic Charts and Reports," quoted in Mukerji (1989, 50).

18. Sverdrup returned to Norway in March 1948 to head the Norsk Polar Institutt (Shor 1978, 34).

19. SIO Archives, Day, "Introduction."

20. Malone, Goldberg, and Munk (1998); Letter, Fox, Hubbs, Shepard, and Zobell to Sproul, May 12, 1950, in oral history interview, Sharp (1985, 47a–47b).

21. Oral history interview, Calvert (1976b).

22. SIO Archives, Letter, Sverdrup to Sproul, March 19, 1943.

23. SIO Archives, Attachment to letter, Sverdrup to Nichols, February 19, 1940.

24. SIO Archives, Letter, Sverdrup to Underhill, February 15, 1945.

25. Oral history interview, Sharp (1985, 64–65).

26. Declassified U.S government documents: Memorandum for the Secretary of Defense, from Aurthur Radford, Chairman, Joint Chiefs of Staff, May 22, 1956; White House, Operations Coordinating Board, Washington, D.C. May 18, 1956.

27. Russian Federation Ministry of Culture. 1999. History of the National Oceanography. Abstracts for the Second International Conference, Kalingrad.

28. Oral history interview, Calvert (1976a); Shor (1978, 404).

29. For a description of the "flagship" model of research planning under the auspices of the International Geosphere Biosphere Program, see Jasanoff and Wynne (1995).

30. For a history of radiation ecology in relation to the discipline of ecology, see Golley (1993, 72, 105).

31. Oral history interview, Sharp (1985, 64); Malone, Goldberg, and Munk (1998, 11).

32. On the IBP, see Golley (1993, 110–140); Jasanoff and Wynne (1995, 54–56).

Chapter 6

1. From Scheiber (1984), p. 419.

2. SIO Archives, Letter, Schaefer to Revelle October 19, 1950. Revelle wrote to Sproul on November 7, 1950, asking what steps should be taken to develop a collaborative program between the IATTC and SIO. Revelle wrote to Schaefer on November 9, 1950, that he could not "imagine a happier relationship" than that which he anticipated between the IATTC and Scripps.

3. It was not the first. See Smith (1994); Mills (1989); Walsh (1999).

4. Convention for the Establishment of an Inter-American Tropical Tuna Commission, Washington, D.C., 1949. Entered into Force March 1950. 1 UST 230, TIAS 2044.

5. Convention for the Establishment of an Inter-American Tropical Tuna Commission, Washington, 1949, Article II.

6. IATTC, Minutes of the Second Annual Meeting, San Jose, Costa Rica, February 1, 1951. Program Details (Planned for Fiscal 1953), 4.

7. IATTC, Minutes of the Third Annual Meeting, San Diego, California, September 1, 1951, 1.

8. See IATTC, Minutes of Annual Meetings, 1950–1998, for data presented by the Director for consideration by the Commission in plenary session.

9. IATTC, Annual Report (1995, 52–53) describes cohort analysis in more detail. NRC (1997, 59) notes that the natural mortality rate is not well known for most fish stocks and must be estimated.

10. IATTC, Annual Report (1995, 17).

11. NRC (1997, 59). IATTC did not propose regulations for the minimum mesh sizes for the purse seine fishery for tunas. (William Bayliff, personal communication, letter dated April 30, 1998, IATTC reference number 0194-800.)

12. IATTC, Annual Report (1995, 58).

13. IATTC, Annual Report (1995, 59).

14. IATTC, Annual Report (1995, 57–61).

15. IATTC, Minutes of the Eighth Annual Meeting, 1956, 2; IATTC, Minutes of the Ninth Annual Meeting, 1957, 6.

16. IATTC, Summary Minutes of the Thirteenth Annual Meeting, Panama City, February 23–24, 1961, 2.

17. Marine Mammal Commission (1975, 23).

18. IATTC, Summary Minutes of the Thirty-Seventh Annual Meeting, October 1979, 4; Bayliff (2001, 39–41).

19. Background Paper No. 3, Thirty-Sixth Annual Meeting of IATTC, October 1978. Research Program and Budget for FY 1980-81, 7. IATTC Archives, La Jolla, California.

20. Background Paper No. 3, Thirty-Sixth Annual Meeting of IATTC, October 1978. Research Program and Budget for FY 1980-81, 7; "Porpoise-Tuna Investigation," Background Paper No. 6, October 1979. See also IATTC, Annual Report (1979).

21. IATTC, Proposed Tuna-Porpoise Research Program and Budget, October 1977–September 1978, 2.

22. See, for example, IATTC, Annual Report (1998, table 35, 204); IATTC, Annual Report (1998, fig. 76, 154); IATTC, Special Report 213 (2001, table 3, 57).

23. Parker (1999, 25). The observation of the fleet's being "in denial" is attributed to Martin Hall.

24. Interview with Dave Bratten, IATTC staff, November 15, 2001. See also Parker (1999, 28).

25. Robin Allen, personal communication, 2001. At this writing, the U.S. sanctions have not been lifted (Brian Hallman, personal communication, November 2001).

26. IATTC, Minutes of the Eighth Annual Meeting, 1956, 2; IATTC, Minutes of the Ninth Annual Meeting, 1957, 6.

27. IATTC, Summary Minutes of the Thirteenth Annual Meeting, February 23–24, 1961, 2. See also Bayliff (2001, 34–38).

28. Interview with Dave Bratten, IATTC staff, November 15, 2001.

29. Interview with Martin Hall, November 1998; Parker (1999, 68–74).

Chapter 7

1. Paragraph added by Sandeep Prasada based on previous discussions with Virginia Walsh.

2. Paragraph added by Sandeep Prasada based on previous discussions with Virginia Walsh.

References

William E. Ritter Papers, Bancroft Library, University of California, Berkeley (Chronological Order)

Diary entry, William E. Ritter, July 2, 1899. Diary 1899. Vol. 1. 71/3 c. Carton 9.

Letters, John Muir to William E. Ritter. October 8, 1899, and October 20, 1899. William E. Ritter Papers. Box 15. Folder Title Muir, John, 1838–1914.

Letters, William E. Ritter to Alexander Agassiz. April 1904, and April 13, 1905.

Letter, William E. Ritter to William James, Harvard University. March 1908.

Letter, John C. Merriam to William E. Ritter. March 31, 1917.

Letter, William E. Ritter to John C. Merriam. October 1919.

Memorandum, National Research Council, Division of Foreign Relations, Committee on Pacific Investigations, p. 1. January 6, 1922.

Letter, William E. Ritter to John C. Merriam. August 3, 1922. William E. Ritter Papers. Folder Title WER Outgoing, 1922.

Letter, William E. Ritter to John C. Merriam. October 9, 1922.

Letter, William E. Ritter to Mrs. Ritter. November 17, 1922.

Letter, William E. Ritter to President David P. Barrows, University of California. April 19, 1923.

Archives, Scripps Institution of Oceanography, La Jolla, California (Chronological Order)

Letter, William E. Ritter to his uncle. William E. Ritter Correspondence, 1889–1890. Folder 38. Box 1. Ritter Family Papers MC 15.

Letter, William E. Ritter to his Uncle Nelson. June 12, 1898. William E. Ritter Correspondence, 1898–ca. 1900. Folder 45. Box 1. Ritter Family Papers MC 15.

Report of William E. Ritter to the Marine Biological Association of San Diego. "Biological Station, 1903." Folder 1. Box 1. Scripps Family Papers 92-38.

Letter, Charles Kofoid to Samons Fletcher Investment Company, San Diego. August 1906. William E. Ritter Correspondence, 1906–1907. Folder 3. Box 1. Kofoid Papers 82-71.

Letter, William E. Ritter to Ellen Scripps. September 15, 1907. "Biological Station, 1907." Folder 5. Box 1. Scripps Family Papers 92-38.

Letter, Charles Kofoid to William E. Ritter. September 29, 1908. Correspondence, September–October 1908. Folder 5. Box 1. Kofoid Papers 82-71.

Letter, William E. Ritter to G. M. Smith, U.S. Bureau of Fisheries, Washington, D.C. July 23, 1917. Signed by Kofoid in Ritter's absence. Correspondence, July 1917. Folder 12. Box 1. Kofoid Papers 82-17.

Letter, T. Wayland Vaughan to R. G. Sproul, October 20, 1924. Correspondence, July–October 1924. Office of the Director (Vaughan) AC 11.

Letter, V. Bjerknes to T. Wayland Vaughan. March 28, 1926. Correspondence, March 1926. Office of the Director (Vaughan) AC 11.

Letter, T. Wayland Vaughan to R. G. Sproul, February 6, 1934. Correspondence, January–February 1934. Office of the Director (Vaughan), AC 11.

Letter, T. Wayland Vaughan to President R. G. Sproul, University of California. June 5, 1934. Correspondence, June 1934. Office of the Director (Vaughan) AC 11.

Letter, T. Wayland Vaughan to Professor George D. Louderback, University of California, Berkeley. December 1935. Correspondence, December 1935. Office of the Director (Vaughan) AC 11.

Letter, Roger Revelle to Harald Sverdrup. October 26, 1936. Correspondence, August–December 1936. Folder 54. Box 1. Roger Revelle Papers MC 6.

Letter, Harald Sverdrup to Vice President Monroe Deutsch, University of California, Berkeley. August 13, 1938. Sverdrup Correspondence, 1938 [July–December]. Folder 12. Office of the Director (Sverdrup) 82-56.

Letter, Harald Sverdrup to L. A. Nichols, Comptroller, University of California, Berkeley. February 19, 1940. Sverdrup Correspondence, 1940 [January–June]. Office of the Director (Sverdrup) 82-56.

Attachment to Letter, Harald Sverdrup to L. A. Nichols, Comptroller, University of California, Berkeley. February 19, 1940. Sverdrup Correspondence, 1940 [January–June]. Office of the Director (Sverdrup) 82-56.

Letter, Harald Sverdrup to R. G. Sproul. June 15, 1942. Sverdrup Correspondence 1942 [January–June]. Office of the Director (Sverdrup) 82-56.

Letter, Harald Sverdrup to President R. G. Sproul, University of California, Berkeley. March 19, 1943. Sverdrup Correspondence, 1943 [January–June]. Office of the Director (Sverdrup) 82-56.

Letters, Harald Sverdrup to Roger Revelle. July 3, 1943, July 19, 1943, and July 20, 1943. Sverdrup Correspondence 1943 [July–December]. Office of the Director (Sverdrup) 82-56.

Report, Harald Sverdrup. April 25, 1944. "Report on the Activity of the Scripps Institution of Oceanography, Biennium 1942–44." Office of the Director (Sverdrup) 81-43.

Letter, Harald Sverdrup to Underhill. February 15, 1945. Sverdrup Correspondence 1945 [January–June]. Office of the Director (Sverdrup) 82-56.

Letter, Solberg to Carl Eckart, March 22, 1948. Office of the Director (Revelle) 81-23. Box 1.

Letter, Roger Revelle to Captain Dundas P. Tucker, Commanding Officer and Director, U.S. Navy Electronics Laboratory, San Diego, California. August 25, 1950.

Letter, Milner Schaefer to Roger Revelle. October 19, 1950.

Letter, Roger Revelle to R. G. Sproul, November 7, 1950. Correspondence, July–October 1950. Office of the Director (Revelle) 1930–61. AC 16. Box 1.

Letter, Roger Revelle to Milner Schaefer, November 9, 1950.

Letter, Roger Revelle to President M. T. Harrington. December 29, 1950. Correspondence, December 1950. Office of the Director (Revelle) 1930–61. AC 16. Box 1. Folder 37.

Memorandum, Roger Revelle to Dean Vern O. Knudsen. July 3, 1952. "Status of the Marine Physical Laboratory During the Academic Year 1952–53." Correspondence, May–July 1952. Office of the Director (Revelle) 1930–61. AC 16. Box 1. Folder 52.

Deborah Day. "Scripps Benefactions: The Role of the Scripps Family in the Founding of the Scripps Institution of Oceanography." (Photocopy obtained from the author August 1999)

Deborah Day. "Introduction: A Guide to the Roger Randall Dougan Revelle Papers." (Photocopy obtained from the author August 1999)

Scripps Institution for Biological Research. Annual Report to the President (various years).

Declassified U.S. Government Documents. Declassified Document Retrieval System (DDRS) (Chronological Order)

White House, Operations Coordinating Board, Washington, D.C. May 18, 1956. Dwight D. Eisenhower Library. DDRS Document 1063. Document ID 199303011063. (Secret)

Memorandum for the Secretary of Defense. Arthur Radford, Chairman, Joint Chiefs of Staff. May 22, 1956. DDRS Document 2473. Document ID 198909012473. (Top Secret)

Letter, John E. Hull, Chairman of the President's Board of Consultants on Foreign Intelligence Activities, to President Eisenhower. October 30, 1958. Dwight D. Eisenhower Library Document 2910. (Top Secret; Declassified 2/26/97)

Memorandum of Conference with the President. Major John S. D. Eisenhower. December 22, 1958. Dwight D. Eisenhower Library Document 2911. (Top Secret; Declassified 4/10/97)

Transcripts of Oral History Interviews. Scripps Institution of Oceanography. Oral History Collection. Texas A&M University, College Station, Texas.

Calvert, Robert. 1976a. "Oceanography Project: Martin Johnson." Oral history interview with Martin Johnson.

Calvert, Robert, 1976b. Oral history interview with Roger Revelle.

Sharp, Sarah L. 1985. "Oceanography, Population Resources and the World." Oral history interview with Roger Revelle. Revelle was Director of Scripps Institution of Oceanography, 1951–1964. SIO Reference Ser. 88-20.

Archives, Inter-American Tropical Tuna Commission (IATTC), La Jolla, California.

Inter-American Tropical Tuna Commission. Annual Reports (various years).

"Porpoise-Tuna Investigation." Background Paper No. 6. October 1979. Prepared for the Thirty-Seventh Meeting of IATTC. (Photocopy obtained from IATTC)

Books and Journal Articles

Andresen, Steinar, Tora Skodvin, Arild Underdal, and Jørgen Wettestad. 2000. *Science and Politics in International Environmental Regimes: Between Integrity and Involvement.* Manchester, U.K.: Manchester University Press.

Arrow, Kenneth. 1995. Information Acquisition and the Resolution of Conflict. In *Barriers to Conflict Resolution,* ed. Kenneth Arrow, Robert Mnookin, Lee Ross, Amos Tversky, and Robert Wilson. New York: W. W. Norton.

Axelrod, Robert, and Robert Keohane. 1986. Achieving Cooperation Under Anarchy: Strategies and Institutions. In *Cooperation Under Anarchy,* ed. Kenneth A. Oye. Princeton, N.J.: Princeton University Press.

Barnes, Barry, David Bloor, and John Henry. 1996. *Scientific Knowledge: A Sociological Approach.* Chicago: University of Chicago Press.

Bayliff, Bill. 2001. *Organizations, Functions, and Achievements of the Inter-American Tropical Tuna Commission.* IATTC Special Report 13. La Jolla, Calif.: IATTC.

Benedick, Richard Elliot. 1998. *Ozone Diplomacy: New Directions in Safeguarding the Planet.* Enlarged ed. Cambridge, Mass.: Harvard University Press.

Beverton, J. J. H., and S. J. Holt. 1957. *On the Dynamics of Exploited Fish Populations.* Ministry of Agriculture, Fisheries and Food (U.K.). Fisheries Investigations, Series 2, 19.

Boyle, James. 1996. *Shamans, Software and Spleens: Law and the Construction of the Information Society.* Cambridge, Mass.: Harvard University Press.

Buchler, Justus, ed. 1955. *The Philosophical Writings of Charles Peirce.* New York: Dover.

Bull, Hedley. 1977. *The Anarchical Society: A Study of Order in World Politics.* New York: Columbia University Press.

Burke, William T. 1994. *The New International Law of Fisheries.* Oxford: Clarendon Press.

Caddy, John F. 1983. An Alternative to Equilibrium Theory for the Management of Fisheries. Paper presented at the Food and Agriculture Organization Expert Consultation in the Regulation of Fishing Effort (Fishing Mortality). Rome: FAO.

Caporaso, James A., and David P. Levine. 1992. *Theories of Political Economy.* New York: Cambridge University Press.

Carr, Edward. 1939. *The Twenty Years' Crisis 1919–1939.* New York: Harper and Row.

Carrier, James. 1987. Marine Tenure and Conservation on Papua New Guinea. In *The Question of the Commons,* ed. James M. McCay and Bonnie J. Acheson. Tucson: University of Arizona Press.

Chayes, Abram, and Antonia Chayes. 1995. *The New Sovereignty.* Cambridge, Mass.: Harvard University Press.

Chomsky, Noam. 2000. *New Horizons in the Study of Language and Mind.* Cambridge: Cambridge University Press.

Clark, Colin Whitcomb. 1976. *Mathematical Bioeconomics: The Optimal Management of Renewable Resources.* New York: Wiley.

Cohen, Michael. 1988. *The History of the Sierra Club 1892–1970.* San Francisco: Sierra Club Books.

Collins, Harry M. 1995. Science Studies and Machine Intelligence. In *Handbook of Science and Technology Studies,* ed. Sheila Jasanoff, Gerald Markle, James Peterson, and Trevor Pinch. Thousand Oaks, Calif.: Sage.

Day, Deborah. 1985. *A Guide to the Roger Randall Dougan Revelle Papers, 1928–1979.* Archives of the Scripps Institution of Oceanography, University of California, San Diego, La Jolla, CA.

Day, Deborah. 1999. Bergen West: Or, How Four Scandinavian Geophysicists Found a Home in the New World. In *Historisch-Meereskundliches Jahrbuch* 6: 69–82. Stralsund, Germany: Deutsches Meeresmuseum.

Deacon, Margaret. 1980. Some Aspects of Anglo-American Cooperation in Marine Science, 1660–1914. In *Oceanography: The Past,* ed. Mary Sears and Daniel Merriman. New York: Springer-Verlag.

DeSombre, Elizabeth. 2000. *Domestic Sources of International Environmental Policy: Industry, Environmentalists, and U.S. Power.* Cambridge, Mass.: MIT Press.

Downs, George, David Rocke, and Randolph Siverson. 1986. Arms Races and Cooperation. In *Cooperation under Anarchy,* ed. Kenneth Oye. Princeton, N.J.: Princeton University Press.

Drayton, William. 1998. Knowledge and Empire. In *The Oxford History of the British Empire.* Oxford: Oxford University Press.

Dreyfus, Hubert, and Paul Rabinow. 1982. *Michel Foucault: Beyond Structuralism and Hermeneutics.* Chicago: University of Chicago Press.

Eckstein, Harry. 1975. Case Study and Theory in Political Science. In *Handbook of Political Science,* ed. Fred I. Greenstein and Nelson W. Polsby. Vol. 1: *Political Science: Scope and Theory.* Reading, Mass.: Addison-Wesley.

Edge, David. 1995. Reinventing the Wheel. In *Handbook of Science and Technology Studies,* ed. Sheila Jasanoff, Gerald Markle, James Peterson, and Trevor Pinch. Thousand Oaks, Calif.: Sage.

Edwards, Paul. 1995. From "Impact" to Social Process: Computers in Society and Culture. In *Handbook of Science and Technology Studies,* ed. Sheila Jasanoff, Gerald Markle, James Peterson, and Trevor Pinch. Thousand Oaks, Calif.: Sage.

Etzkowitz, Henry, and Andrew Webster. 1995. Science as Intellectual Property. In *Handbook of Science and Technology Studies,* ed. Sheila Jasanoff, Gerald Markle, James Peterson, and Trevor Pinch. Thousand Oaks, Calif.: Sage.

Falkowski, Lawrence. 1979. *Psychological Models in International Politics.* Boulder, Colo.: Westview Press.

FAO (Food and Agriculture Organization). 1995. *Precautionary Approach to Fisheries.* FAO Fisheries Technical Report 350. Rome: FAO.

———. 1996. *The State of World Fisheries and Aquaculture.* Rome: FAO.

Ferguson, Yale H., and Richard W. Mansbach. 1996. *Polities: Authority, Identities, and Change.* Columbia: University of South Carolina Press.

Fischer, Frank. 2000. *Citizens, Experts and the Environment.* Durham, N.C.: Duke University Press.

Fitzgerald, Deborah. 1990. *The Business of Breeding: Hybrid Corn in Illinois, 1890–1940.* Ithaca: Cornell University Press.

Fox, Stephen. 1986. *The American Conservation Movement: John Muir and His Legacy.* Madison: University of Wisconsin Press.

Friedman, Robert. 1989. *Appropriating the Weather: Vilhelm Bjerknes and the Construction of a Modern Meteorology.* Ithaca, N.Y.: Cornell University Press.

———. 1994. *The Expeditions of Harald Ulrik Sverdrup: Contexts for Shaping an Ocean Science.* La Jolla, Calif.: Scripps Institution of Oceanography.

Gaddis, John L. 1997. *We Now Know: Rethinking Cold War History.* Oxford: Oxford University Press.

Gallie, Walter Bryce. 1952. *Peirce and Pragmatism.* Edinburgh: Penguin Books.

Gifford, Terry, ed. 1996. *John Muir: His Life and Letters and Other Writings.* London: Baton Wicks Publications.

Goldstein, Judith. 1993. *Ideas, Interests, and American Trade Policy*. Ithaca, N.Y.: Cornell University Press.

Goldstein, Judith, and Robert Keohane, eds. 1993. *Ideas and Foreign Policy: Beliefs, Institutions, and Political Change*. Ithaca, N.Y.: Cornell University Press.

Goldstein, Judith, Miles Kahler, Robert Keohane, and Anne-Marie Slaughter. 2000. Introduction: Legalization and World Politics. *International Organization* 54 (3): 385–399.

Golley, Frank. 1993. *A History of the Ecosystem Concept in Ecology*. New Haven, Conn.: Yale University Press.

Gordon, H. Scott. 1954. The Economic Theory of a Common-Property Resource: The Fishery. *Journal of Political Economy* 62: 124–142.

Gould, Stephen Jay. 1977. *Ever Since Darwin: Reflections in Natural History*. New York: W. W. Norton.

Gower, Barry. 1997. *Scientific Method*. New York: Routledge.

Guston, David. 2000. *Between Science and Politics: Assuring the Integrity and Productivity of Research*. Cambridge: Cambridge University Press.

Haas, Peter M. 1990. *Saving the Mediterranean*. Princeton, N.J.: Princeton University Press.

———, ed. 1992. Knowledge, Power, and International Policy Coordination. *International Organization* 46 (1): 1–35.

———. 1997. Scientific Communities and Multiple Paths to Environmental Management. In *Saving the Seas*, ed. L. A. Brooks and S. VanDeveer. Silver Spring: University of Maryland Press.

———. 1998. Compliance with EU Directives: Insights from International Relations and Comparative Politics. *Journal of European Public Policy* 5 (1): 38–65.

———. 1999. Social Constructivism and the Evolution of Multilateral Environmental Governance. In *Globalization and Governance*, ed. Aseem Prakash and Jeffrey Hart. London: Routledge.

Haas, Peter M., Robert Keohane, and Marc Levy, eds. 1993. *Institutions for the Earth*. Cambridge, Mass.: MIT Press.

Haas, Peter M., and David McCabe. 2001. Amplifiers or Dampeners: International Institutions and Social Learning in the Management of Global Environmental Risks. In *Learning to Manage Global Environmental Risks*. Vol. 1. The Social Learning Group. Cambridge, Mass.: MIT Press.

Hackmann, Willem. 1984. *Seek and Strike: Sonar, Anti-Submarine Warfare, and the Royal Navy 1914–1954*. London: HMSO.

———. 1986. Sonar Research and Naval Warfare 1914–1954: A Case Study of Twentieth-Century Establishment Science. *Historical Studies in the Physical and Biological Sciences* 16 (1): 83–110.

Hardin, Garrett. 1968. The Tragedy of the Commons. *Science* 162: 1243–1248.

Hart, David M., and David G. Victor. 1993. Scientific Elites and the Making of U.S. Policy for Climate Change Research 1957–1974. *Social Studies of Science* 23: 643–680.

Hasenclever, Andreas, Peter Mayer, and Volker Rittberger. 1996. Interests, Power, Knowledge: The Study of International Regimes. *Mershon International Studies Review* 40: 177–228.

Haslam, S. Alexander, Craig McGarty, John C. Turner, Judith Nye, and Aaron Brower, eds. 1996. Salient Group Memberships and Persuasion. In *What's Social about Social Cognition?* Thousand Oaks, Calif.: Sage.

Hayek, Friedrich A. 1948. Economics and Knowledge. In *Individualism and Economic Order*. Chicago: University of Chicago Press.

Hilborn, Ray, and John Sibert. 1988. Is International Tuna Management Necessary? *Marine Policy* (January): 31–39.

Hillman, H. C. 1952. Comparative Strength of the Great Powers. In *The World in March 1939,* ed. Arnold Toynbee. New York: Oxford University Press.

Hinsley, Francis H. 1986. In *Sovereignty*. 2d ed. New York: Cambridge University Press.

Hjort, Johan. 1933. Whales and Whaling. *Hvalradets Skrifter* 7: 7–29.

Hobbes, Thomas. 1651. *Leviathan*. New York: Collier Books.

Holland, John H. 1995. *Hidden Order: How Adaptation Builds Complexity*. Reading, Mass.: Perseus Books.

Hollick, Ann L. 1981. *U.S. Foreign Policy and the Law of the Sea*. Princeton, N.J.: Princeton University Press.

Holloway, David. 1994. *Stalin and the Bomb: The Soviet Union and Atomic Energy 1939–1956*. New Haven, Conn.: Yale University Press.

Hønneland, Geir. 2000. *Coercive and Discursive Compliance Mechanisms in the Management of Natural Resources*. Boston: Kluwer.

Houghton, John T., L. G. Meira Filho, B. A. Callander, N. Harris, A. Kattenberg, and K. Maskell, eds. 1995. *Climate Change 1995: The Science of Climate Change*. Cambridge: Cambridge University Press.

Hull, David L. 1988. *Science as a Process: An Evolutionary Account of the Social and Conceptual Development of Science*. Chicago: University of Chicago Press.

Indo-Pacific Fishery Commission. 1994. *Socio-Economic Issues in Coastal Fisheries Management,* Bangkok: Regional Office for Asia and the Pacific, Food and Agriculture Organization.

Jasanoff, Sheila. 1990. *The Fifth Branch*. Cambridge, Mass.: Harvard University Press.

———. 1995. *Science at the Bar.* Cambridge, Mass.: Harvard University Press.

———. 1997. Compelling Knowledge in Public Decisions. In *Saving the Seas,* ed. L. A. Brooks and S. VanDeveer. Silver Spring: University of Maryland Press.

———. 1998. Contingent Knowledge: Implications for Implementation and Compliance. In *Engaging Countries,* ed. Edith Brown Weiss and Harold K. Jacobson. Cambridge, Mass.: MIT Press.

Jasanoff, Sheila, Gerald Markle, James Peterson, and Trevor Pinch, eds. 1995. *Handbook of Science and Technology Studies*. Rev. ed. Thousand Oaks, Calif.: Sage.

Jasanoff, Sheila, and Brian Wynne. 1995. Science and Decision Making. In *Human Choice and Climate Change,* ed. Steve Rayner and Elizabeth Malone. Vol. 1. Columbus, Ohio: Battelle Press.

Jervis, Robert. 1976. *Perception and Misperception in International Politics*. Princeton, N.J.: Princeton University Press.

Johannes, Robert Earle. 1981. *Words of the Lagoon: Fishing and Marine Lore in the Palau District of Micronesia*. Berkeley: University of California Press.

Joseph, James. 1994. The Tuna-Dolphin Controversy in the Eastern Pacific Ocean: Biological, Economic, and Political Impacts. *Ocean Development and International Law* 25 (1): 5.

Keegan, John. 1999. *The First World War*. New York: Knopf.

Kennedy, Paul. 1988. *The Rise and Fall of the Great Powers*. New York: Vintage Books.

Keohane, Robert. 1982. The Demand for International Regimes. *International Organization* 36 (2): 325–355.

———. 1984. *After Hegemony*. Princeton, N.J.: Princeton University Press.

———. 1988. International Institutions: Two Approaches. *International Studies Quarterly* 32: 379–396.

Keohane, Robert, ed. 1986. *Neorealism and Its Critics*. New York: Columbia University Press.

Keohane, Robert, Andrew Moravcsik, and Anne-Marie Slaughter. 2000. Legalized Dispute Resolution: Interstate and Transnational. *International Organization* 54 (3): 457–488.

Keohane, Robert, and Joseph S. Nye. 1989. *Power and Interdependence*. 2d ed. New York: HarperCollins.

Kevles, Daniel J. 1971. *The Physicists: History of a Scientific Community in Modern America*. Cambridge, Mass.: Harvard University Press.

Kim, Woosang, and Bruce Bueno de Mesquita. 1995. How Perceptions Influence the Risk of War. *International Studies Quarterly* 39: 51–65.

King, Gary, Robert Keohane, and Sidney Verba. 1994. *Designing Social Inquiry*. Princeton, N.J.: Princeton University Press.

Kohler, Robert. 1982. *From Medical Chemistry to Biochemistry: The Making of a Biomedical Discipline*. Cambridge: Cambridge University Press.

———. 1994. *Lords of the Fly: Drosophila Genetics and the Experimental Life*. Chicago: University of Chicago Press.

Krasner, Stephen D. 1999. *Sovereignty: Organized Hypocrisy*. Princeton, N.J.: Princeton University Press.

Krasner, Stephen D., ed. 1983. *International Regimes*. Ithaca, N.Y.: Cornell University Press.

Kratochwil, Friedrich. 1989. *Rules, Norms, and Decisions*. Cambridge: Cambridge University Press.

———. 1995. Sovereignty as Dominium. In *Beyond Westphalia? National Sovereignty and International Intervention,* ed. Gene M. Lyons and Michael Mastanduno. Baltimore: Johns Hopkins University Press.

Lasky, Martin. 1975. Historical Review of Underwater Acoustic Technology, 1939–1945, with an Emphasis on Undersea Warfare. *U.S. Navy Journal of Underwater Acoustics* 25 (October): 885–918.

Latour, Bruno, and Steve Woolgar. 1979. *Laboratory Life: The Social Construction of Scientific Facts*. Beverly Hills, Calif.: Sage.

Lessig, Lawrence. 1999. *Code and Other Laws of Cyberspace*. New York: Basic Books.

Litfin, Karen. 1994. *Ozone Discourses: Science and Politics in Global Environmental Cooperation*. New York: Columbia University Press.

———. 1998. *The Greening of Sovereignty in World Politics*. Cambridge, Mass.: MIT Press.

Ludwig, Donald, Ray Hilborn, and C. Walters. 1993. Uncertainty, Resource Exploitation, and Conservation: Lessons from History. *Science* 260: 17.

MacLeish, William H. 1982. Profile: Roger Revelle, Senior Senator of Science. *Oceanus* 25 (Winter): 67–70.

MacLeod, Roy, ed. 2000. *Science and the Pacific War: Science and Survival in the Pacific 1939–1945*. Boston: Kluwer.

MacLeod, Roy, and Philip Rehbock, eds. 1988. *Nature in Its Greatest Extent: Western Science in the Pacific*. Honolulu: University of Hawaii Press.

Makhijani, Arjun, and Kevin Gurney. 1995. *Mending the Ozone Hole: Science, Technology, and Policy*. Cambridge, Mass.: MIT Press.

Malone, Thomas F., Edward D. Goldberg, and Walter H. Munk. 1998. *Roger Randall Dougan Revelle 1901–1991*. Biographical Memoirs. Vol. 75. Washington, D.C.: National Academy Press.

Marashi, S. H. 1996. *Summary Information on the Role of International Fishery and Other Bodies with Regard to the Conservation and Management of Living Resources of the High Seas*. Fisheries Circular 908. Rome: FAO.

March, James, and Johan Olsen. 1989. *Rediscovering Institutions: The Organizational Basis of Politics*. New York: Free Press.

———. 1998. Institutional Dynamics of International Political Orders. *International Organization* 52 (4): 943–969.

May, Ernst. 1959. *The World War and American Isolation 1914–1917*. Cambridge, Mass.: Harvard University Press.

McCay, Bonnie. 1996. Common and Private Concerns. In *Rights to Nature: Ecological, Economic, Cultural, and Political Principles of Institutions for the Environment,* ed. Susan S. Hanna, Carl Folke, and Karl-Goran Maler. Washington, D.C.: Island Press.

McCay, Bonnie, and James M. Acheson. 1987. *The Question of the Commons: The Culture and Ecology of Communal Resources*. Tucson: University of Arizona Press.

McEwen, George F., and Ellis Michael. 1913–1916. Hydrographic, Plankton, and Dredging Records of the Scripps Institution for Biological Research of the University of California, 1901–1912 [and Continuation, 1913–1915]. *University of California Publications in Zoology* 15: 1–254.

Mearsheimer, John. 1994. The False Promise of International Institutions. *International Security* 19 (3): 5–49.

Mills, Eric L. 1980. Alexander Agassiz, Carl Chun and the Problem of the Intermediate Fauna. In *Oceanography: The Past*, ed. Mary Sears and Daniel Merriman. New York: Springer-Verlag.

———. 1989. *Biological Oceanography: An Early History, 1870–1960*. Ithaca, N.Y.: Cornell University Press.

———. 1991. The Oceanography of the Pacific: George F. McEwen, H. U. Sverdrup, and the Origin of Physical Oceanography on the West Coast of North America. *Annals of Science* 48: 241–266.

———. 1993. *The Scripps Institution of Oceanography: Origin of a Habitat for Ocean Science*. San Diego, Calif.: Scripps Institution of Oceanography.

Milner, Helen. 1997. *Interests, Institutions, and Information*. Princeton, N.J.: Princeton University Press.

Mitchell, Ronald. 1994. *Intentional Oil Pollution at Sea*. Cambridge, Mass.: MIT Press.

———. 1998. Sources of Transparency: Information Systems in International Regimes. *International Studies Quarterly* 42 (March): 109–130.

Modelski, George, and William Thompson. 1988. *Seapower in Global Politics 1494–1993*. Seattle: University of Washington Press.

Moravcsik, Andrew. 1997. Taking Preferences Seriously: A Liberal Theory of International Politics. *International Organization* 51 (4): 513–553.

Morgan, Neil, and Judith Morgan. 1996. *Roger: A Biography of Roger Revelle*. San Diego, Calif.: Scripps Institution of Oceanography.

Muir, John. 1911. *My First Summer in the Sierra*. Boston: Houghton Mifflin; Riverside Press.

Mukerji, Chandra. 1989. *A Fragile Power: Scientists and the State*. Princeton, N.J.: Princeton University Press.

Munk, Walter. 1991. The Revelle Years. In *Scripps Institution of Oceanography Annual Report*. San Diego, Calif.: Scripps Institution of Oceanography.

Murray, John. 1911. Alexander Agassiz: His Life and Work. *Science* 33 (June 9): 873–887.

NAS (National Academy of Sciences). 1976. *Marine Scientific Research and the Third Law of the Sea Conference*. Washington, D.C.: National Academy of Sciences, 1976.

NAS/NRC (National Academy of Sciences/National Research Council). 1957. *The Effects of Atomic Radiation on Oceanography and Fisheries: Report of the Committee on Effects of Atomic Radiation on Oceanography and Fisheries of the National Academy of Sciences Study of the Biological Effects of Atomic Radiation.* Washington, D.C.: Publication 551.

Nash, Roderick. 1982. *Wilderness and the American Mind.* New Haven, Conn.: Yale University Press.

Nierenberg, William. 1996. *Harald Ulrik Sverdrup 1888–1957.* Biographical Memoirs 69. Washington, D.C.: National Academy Press.

Nornvall, Frederik. 1999. Reasons for Marine Science: Arguments for Hydrographic Research in Sweden. *Historisch-Meereskundliches Jahrbuch.* Stralsund, Germany: Deutsches Meeresmuseum.

Norse, Elliott A., ed. 1993. *Global Marine Biological Diversity.* Washington, D.C.: Island Press.

North, Douglass C. 1981. *Structure and Change in Economic History.* New York: W. W. Norton.

NRC (National Research Council). 1992. *Dolphins and the Tuna Industry.* Washington, D.C.: National Academy Press.

———. 1995. *Understanding Marine Biodiversity.* Washington, D.C.: National Academy Press.

———. 1997. *Improving Fish Stock Assessments.* Washington, D.C.: National Academy Press.

Odell, John. 1982. *U.S. International Monetary Policy: Markets, Power and Ideas as Sources of Change.* Princeton, N.J.: Princeton University Press.

Olson, Mancur. 1965. *The Logic of Collective Action.* Cambridge, Mass.: Harvard University Press.

Oreskes, Naomi, and Ronald Rainger. 2000. Science and Security Before the Atomic Bomb: The Loyalty Case of Harald U. Sverdrup. *Studies in History and Philosophy of Science Part B: Studies in History and Philosophy of Modern Physics* 31 (3): 309–369.

Ostrom, Elinor. 1990. *Governing the Commons: The Evolution of Institutions for Collective Action.* Cambridge: Cambridge University Press.

Ostrom, Elinor, J. Burger, C. B. Field, R. B. Norgaard, and D. Policansky. 1999. Revisiting the Commons: Local Lessons, Global Challenges. *Science* 284: 278–282.

Oye, Kenneth, ed. 1986. *Cooperation under Anarchy.* Princeton, N.J.: Princeton University Press.

Parker, Richard. 1999. The Use and Abuse of Trade Leverage to Protect Global Commons: What We Can Learn from the Tuna-Dolphin Conflict. *Georgetown International Environmental Law Review* 12 (1).

Peirce, Charles S. 1877. The Fixation of Belief. In *The Philosophical Writings of Charles S. Pierce,* ed. Justus Buchler. New York: Dover, 1955.

Polanyi, Michael. 1956. *Personal Knowledge*. Chicago: University of Chicago Press.

Popper, Karl. 1972. *Objective Knowledge: An Evolutionary Approach*. Oxford: Oxford University Press.

Prager, Frank D. 1944. A History of Intellectual Property from 1545 to 1787. *Journal of the Patent Office Society* 26 (11): 711–760.

Rainger, Ronald. 2000. Science at the Crossroads: The Navy, Bikini Atoll, and American Oceanography in the 1940s. *Historical Studies in the Physical and Biological Sciences* 30 (pt. 2): 349–371.

Raitt, Helen, and Beatrice Moulton. 1967. *Scripps Institution of Oceanography: First Fifty Years*. Los Angeles: Ward Ritchie Press.

Rayner, Steve, and Elizabeth Malone, eds. 1988. *Human Choice and Climate Change*. Columbus, Ohio: Battelle Press.

Revelle, Roger. 1954. Foreword. In *Geology of Bikini and Nearby Atolls*, by Kenneth O. Emery, J. I. Tracey, Jr., and H. S. Ladd. Washington, D.C.: Government Printing Office.

Revelle, Roger. 1969. The Age of Innocence and War in Oceanography. *Oceans* 1 (3): 6–16.

Rhodes, Richard. 1986. *The Making of the Atomic Bomb*. New York: Touchstone.

———. 1995. *Dark Sun: The Making of the Hydrogen Bomb*. New York: Simon and Schuster.

Ritter, Mary. 1933. *More Than Gold in California 1849–1933*. Berkeley, Calif.: Professional Press.

Ritter, William E. 1899. The Harriman Alaska Expedition. *The University Chronicle* (August). Archives, Scripps Institution of Oceanography, La Jolla, Calif.

———. 1906. Octacnemus: Report on the Scientific Results of the Expedition to the Eastern Tropical Pacific, in Charge of Alexander Agassiz, by the U.S. Fish Commission Steam *Albatross*, from October 1904 to March 1905, Lieutenant Commander L. M. Garrett, USN, Commanding. *Bulletin of the Museum of Comparative Zoology* 46 (13).

———. 1908. The Scientific Work of the San Diego Marine Biological Station in the Year 1908. *Science* N.S. 28 (715).

———. 1911. The Meaning of Vitalism. *Science* N.S. 33 (851).

———. 1915. *War, Science, and Civilization*. Boston: Sherman French.

———. 1919. *An Organismal Theory of Consciousness*. Boston: Gorham Press.

———. 1927. *The Natural History of Our Conduct*. New York: Harcourt, Brace.

Rozwadowski, Helen M. 1999. Marine Science in the Age of Internationalism. *Historisch-Meereskundliches Jahrbuch*. Stralsund, Germany: Deutsches Meeresmuseum.

Ruddle, Kenneth, and Robert E. Johannes, eds. 1985. The Traditional Knowledge and Management of Coastal Systems in Asia and the Pacific. Paper presented at a regional seminar, UNESCO Regional Office for Science and Technology for Southeast Asia, Jakarta, December 5–9, 1983.

Ruggie, John. 1983. International Regimes, Transactions, and Change: Embedded Liberalism in the Postwar Economic Order. In *International Regimes*, ed. Stephen D. Krasner, 195–231. Ithaca, N.Y.: Cornell University Press.

———. 1993. Territoriality and Beyond. *International Organization* 47 (1): 139–174.

———. 1998. What Makes the World Hang Together? Neo-Utilitarianism and the Social Constructivist Challenge. *International Organization* 52 (4): 855–885.

Scheiber, Harry. 1984. U.S. Pacific Fishery Studies 1945–1970: Oceanography, Geopolitics, and Marine Fisheries Expansion. In *Ocean Sciences: Proceedings of the 4th International Congress on the History of Oceanography*, ed. Walter Lenz and Margaret Deacon.

———. 1988. Wilbert Chapman and the Revolution in U.S. Pacific Ocean Science and Policy. In *Nature in Its Greatest Extent: Western Science in the Pacific*, ed. Roy MacLeod and Philip Rehbock. Honolulu: University of Hawaii Press.

Schlee, Susan. 1973. *The Edge of an Unfamiliar World: A History of Oceanography*. New York: Dutton.

Seaborg, Glenn T. 1983. *Kennedy, Khrushchev, and the Test Ban*. Berkeley: University of California Press.

Searle, John. 1995. *The Construction of Social Reality*. New York: Free Press.

Sears, Mary, and Daniel Merriman, eds. 1980. *Oceanography: The Past*. New York: Springer-Verlag.

Shapin, Steven. 1994. *A Social History of Truth*. Chicago: University of Chicago Press.

———. 1996. *The Scientific Revolution*. Chicago: University of Chicago Press.

Shor, Elizabeth Noble. 1978. Scripps Institution of Oceanography: Probing the Oceans 1936–1976. San Diego, California: Tofua Press.

Smit, Wim A. 1995. Science, Technology, and the Military. In *Handbook of Science and Technology Studies*, ed. Sheila Jasanoff, Gerald Markle, James Peterson, and Trevor Pinch. Thousand Oaks, Calif.: Sage.

Smith, Tim. 1994. *Scaling Fisheries*. Cambridge: Cambridge University Press.

Social Learning Group. 2001. *Learning to Manage Global Environmental Risks*. 2 vols. Cambridge, Mass.: MIT Press.

Sontag, Sherry, and Christopher Drew. 1998. *Blind Man's Bluff: The Untold Story of American Submarine Espionage*. New York: Public Affairs.

Sprout, Harold, and Margaret Sprout. 1942. *The Rise of American Naval Power 1776–1918*. Princeton, N.J.: Princeton University Press.

———. 1943. *Toward A New Order of Sea Power: American Naval Policy and the World Scene 1918–1922*. New York: Greenwood Press.

Spruyt, Hendrik. 1994. *The Sovereign State and Its Competitors*. Princeton, N.J.: Princeton University Press.

Thompson, W. F., and H. Bell. 1934. *Biological Statistics of the Pacific Halibut Fishery*. Vancouver: Wrigley Printing Co.

Toynbee, Arnold, ed. 1952. *The World in March 1939*. New York: Oxford University Press.

Tuomela, Raimo. 1995. *The Importance of Us*. Princeton, N.J.: Princeton University Press.

———. 2000a. *Cooperation*. Boston: Kluwer.

———. 2000b. Belief Versus Acceptance. *Philosophical Explorations* 3 (2): 122–137.

———. 2002. Collective Intentionality, Organizations, and Institutions. Third Conference on Collective Intentionality, Rotterdam, December 12–14.

Turekian, Karl. 1996. *Global Environmental Change: Past, Present and Future*. Upper Saddle River, N.J.: Prentice-Hall.

Turner, John C. 1985. Social Categorization and the Self-Concept: A Social Cognitive Theory of Group Behavior. *Advances in Group Processes* 2:77–121.

Turner, John C., Penelope J. Oakes, S. Alexander Haslam, and Craig McGarty. 1994. Self and Collective: Cognition and Social Context. *Personality and Social Psychology Bulletin* 20 (5): 454–463.

Vattel, Emer de. 1805. *The Law of Nations; or, Principles of the Law of Nature Applied to the Conduct and Affairs of Nations and Sovereigns. A Work Tending to Display the True Interest of Powers*. Northampton, Mass.: Printed by Thomas M. Pomroy, for S. & E. Butler.

Victor, David G., Kal Raustiala, and Eugene Skolnikoff. 1998. *The Implementation and Effectiveness of International Environmental Commitments*. Cambridge, Mass.: MIT Press.

Victor, David G. 2001. *The Collapse of the Kyoto Protocol and the Struggle to Slow Global Warming*. Princeton, N.J.: Princeton University Press.

Walker, Robert. 1993. *Inside/Outside: International Relations as Political Theory*. Cambridge: Cambridge University Press.

Walsh, Virginia. 1998. Eliminating Driftnets from the North Pacific Ocean. *Ocean Development and International Law* 29: 295–322.

———. 1999. Illegal Whaling for Humpbacks by the Soviet Union in the Antarctic, 1947–1972. *Journal of Environment and Development* 8 (3): 307–327.

Waltz, Kenneth. 1979. *Theory of International Politics*. New York: McGraw-Hill.

Weber, Max. 1913/1990. The Social Psychology of the World's Religions. In *From Max Weber: Essays in Sociology*, trans. H. H. Gerth and C. W. Mills. New York: Oxford University Press.

Webster, A. 1990. Institutional Stability: Engineering an Environment for Biotechnology. *Science and Public Policy* (17) 5.

Weisgall, Jonathan M. 1994. *Operation Crossroads: The Atomic Tests at Bikini Atoll.* Annapolis, Md.: Naval Institute Press.

Wendt, Alexander. 1992. Anarchy Is What States Make of It. *International Organization* 46: 391–425.

Wendt, Alexander. 1999. *Social Theory of International Politics.* Cambridge: Cambridge University Press.

Wight, Martin. 1952. The Balance of Power. In *The World in March 1939,* London: Oxford University Press.

Wüst, Georg. 1964. Deep Sea Expeditions and Research Vessels 1873–1960. In *Progress in Oceanography,* ed. M. Sears. Vol. 2. New York: Pergamon Press.

Yarbrough, Beth V., and Robert M. Yarbrough. 1990. International Institutions and the New Economics of Organization. *International Organization* 44: 235–259.

Yee, Albert S. 1996. The Causal Effects of Ideas on Policies. *International Organization* 50 (1): 69–108.

York, Herbert F. 1976. *The Advisors: Oppenheimer, Teller and the Superbomb.* Palo Alto, Calif.: Stanford University Press.

Young, Oran. 1994. *International Governance: Protecting the Environment in a Stateless Society.* Ithaca, N.Y.: Cornell University Press.

———. 1996. Rights, Rules, and Resources in International Society. In *Rights to Nature,* ed. Susan S. Hanna, Carl Folke, and K. G. Maler. Washington, D.C.: Island Press.

———. 1999. *Governance in World Affairs.* Ithaca, N.Y.: Cornell University Press.

———. 2001. The Behavioral Effects of Environmental Regimes: Collective Action vs. Social Practice Models. *International Environmental Agreements: Law, Politics and Economics* 1: 9–29.

Young, Oran, ed. 1999. *Science Plan.* Bonn: International Human Dimensions Programme (IHDP).

Zubok, Vladislav, and Constantine Pleshakov. 1996. *Inside the Kremlin's Cold War.* Cambridge, Mass.: Harvard University Press.

Index

Printed in the United States
by Baker & Taylor Publisher Services